Semiconducting Phosphorus

Emerging Materials and Technologies
Editor Boris I. Kharissov

Semiconducting Black Phosphorus

From 2D Nanomaterial to Emerging 3D Architecture

Han Zhang, Nasir Mahmood Abbasi and Bing Wang

CRC Press
Taylor & Francis Group
Boca Raton London New York

CRC Press is an imprint of the
Taylor & Francis Group, an **informa** business

First edition published 2022
by CRC Press
6000 Broken Sound Parkway NW, Suite 300, Boca Raton, FL 33487-2742

and by CRC Press
2 Park Square, Milton Park, Abingdon, Oxon OX14 4RN

Library of Congress Cataloging-in-Publication Data
A catalog record has been requested for this book

ISBN: 978-1-032-06763-6 (hbk)
ISBN: 978-1-032-10805-6 (pbk)
ISBN: 978-1-003-21714-5 (ebk)

DOI: 10.1201/9781003217145

Typeset in Times
by Newgen Publishing UK

Contents

Preface

Black phosphorus (BP) is a rising star among two-dimensional materials other than graphene due to its unique physical and chemical properties. The two-dimensional (2D) BP can be formed into a three-dimensional (3D) form (such as a macrostructure and a hybrid structure) in a highly complex manner to better provide better practical application performance. Therefore, people need to study and improve the 3D structure of BP to manufacture functional nanodevices with customized features and characteristics. BP's 3D hybrid system has huge van der Waals (vdWs) interactions with other 2D layered materials, quantum dots, polymers, and small organic molecules, and has a good host/hybrid structure, and has enhanced electronic, mechanical, and biodegradable characteristics, etc.

In addition, the 3D macrostructure of two-dimensional materials is porous, so they can be used as electrodes for battery and supercapacitor applications. Moreover, the pores can accommodate various other materials, including metals, organic molecules, polymers, etc., resulting in new functions and applications. This book summarizes some information about 2D BP monolayer composition and its scalable composition, characteristics, and assembly into a 3D architecture. Also discussed in detail is the interaction of BP two-dimensional nanomaterials with other materials (such as polymers, organic molecules, other inorganic materials), as well as with other dihydrocarbons (TMD graphene and boron nitride) and other 3D architectures. The BP structure for comparative research strategies is also examined. Finally, the application of BP's 3D macrostructure (formed by functionalization on 2D BP) in various fields such as energy, biomedicine, and catalysis is reviewed.

Authors' Biographies

Prof. Dr. Han Zhang obtained his B.S. degree from Wuhan University in 2006 and Ph.D. from Nanyang Technological University in 2010. He is currently a Director of the Shenzhen Key Laboratory of 2D Materials and Devices, and the Shenzhen Engineering Laboratory of Phosphorene and Optoelectronics, Shenzhen University. His current research focus is the ultrafast and nonlinear photonics of 2D materials.

Dr. Nasir Mahmood Abbasi received his Ph.D. in Chemical Engineering with a major research direction in the chemistry and physics of nanomaterials from Zhejiang University, China in 2016. He worked as a postdoc fellow in the School of Physics and Optoelectronic Engineering at Shenzhen University with a major in nanomaterials during 2019–2021. His research interests include heterostructures and the macroassemblies of 2D materials for their applications in energy, healthcare, and electronics.

Prof. Dr. Bing Wang received her Ph.D. degree in physics from Sun Yat-Sen University in 2007. Currently, she is associate professor in the College of Physics and Optoelectronic Engineering, Shenzhen University. Her scientific research concentrates on 2D semiconductors' optoelectronic properties and applications. She has been selected as one of the most outstanding young teachers in Guangdong province and won the second prize of the National Natural Science Foundation of China.

1 General Overview of 2D Black Phosphorus

1.1 INTRODUCTION

Investigating unknown phases of materials can inspire new fields of exploration for scientific communities, as both experimentalists and theorists. Due to their unique properties in electronics, photonics, and biomedicine, the low dimensional systems (zero-, one-, or two-dimension) have been gaining increasing attention in recent years. Different varieties of 2D nanomaterial are in the developing stage (Neto et al. 2009; Kis 2012), with their unique characteristics differentiating them from other nanomaterial such as zero- and one-dimensional nanomaterial in the case of fundamental research as well as in the field of applications. For example, a 2D material graphene has outstanding mechanical, electronic, and thermal properties, and 2D transition metal dichalcogenides (TMDCs) such as MoS_2, $MoSe_2$, WS_2, and WSe_2 have peculiar optical properties (Splendiani et al. 2010; Wang, Kalantar-Zadeh, et al. 2012) and a series of favorable uses in the fields of catalysis, optoelectronics, sensing (chemical and biological sensors), supercapacitors, and energy storage devices (Xu et al. 2013; Kong et al. 2016; Liu, Zhao, et al. 2014; Wang, Chen, et al. 2012; Zhao et al. 2007; 2008a, 2008b, 2009a, 2009b, 2010; Chen et al. 2013, 2014; Zhang et al. 2008, 2009, 2011; Li et al. 2016; Yan et al. 2014; Shao et al. 2015; Wang et al. 2014; Liu et al. 2015; Song, Zhang, and Tang 2012; Wang, Chen, et al. 2015; Mao, Jiang, et al. 2015; Ren et al. 2017; Lu et al. 2013; Guo et al. 2017; Wang et al. 2017; Britnell et al. 2012; Wan et al. 2015; Ma et al. 2015). The literature has confirmed the importance of 2D materials and highlighted the inauguration of the investigation of 2D nanomaterials by the successful exfoliation of graphene (Novoselov et al. 2005; Neto and Novoselov 2011; Butler et al. 2013; Kong et al. 2016).

Despite the zero bandgaps, the planer structure and the semi-metallic characteristics of graphene and the relatively low mobility value in TMDCs limit their practical applications in the field of electronics and photonics (Xiang et al. 2014; Ponraj et al. 2016; Mu et al. 2015; Koenig et al. 2014; Liu, Neal, et al. 2014). So the development of new 2D layered material to overcome these problems is highly desired (Koenig et al. 2014; Liu, Neal, et al. 2014). Recently black phosphorus (BP) has shown a remarkable advancement in series of novel possessions, such as a tunable and narrow bandgap as well as high-carrier mobility (Rahman et al. 2016; Mao, Tang, et al. 2015; Li et al. 2014; Liu, Neal, et al. 2014; Castellanos-Gomez et al. 2014; Xia, Wang, and

DOI: 10.1201/9781003217145-1

Jia 2014; Rudenko and Katsnelson 2014; Qiao et al. 2014; Wang, Jones, et al. 2015), a high efficiency of photothermal transformation (Shao et al. 2016; Sun et al. 2015), high chemical reactivity that might be responsible for a high amount of drug-packing capability and as a result increase the capacity of loading (Chen et al. 2017), and excellent biodegradation and biocompatibility (Shao et al. 2016). Due to these multi-directional properties, BP has been synthesized and investigated as an active research area for scientific communities during the past five years. Like a puckered structure, its geometry supports the tuning of a band-like structure due to the vertical field. It has high charge-carrier mobility (at room temperature, bulk BP exhibits electron and hole mobilities of 220 and 350 cm^2 V^{-1} s^{-1}), narrow and shortest bandgap which deviated from 0.3 eV (in bulk) to 1.5 eV (in monolayer), and exceptional anisotropic structure (on same plane) as a growing star (Lei et al. 2017; Li et al. 2014). 2D BP has a series of practical applications and has recently become an active area of research curiosity in the areas of material science, condensed matter physics, and chemistry. Owing to its puckered monolayer unique geometry having many unprecedented properties, 2D BP was employed for various applications such as nano/microelectronics, biomedicines, and chemical sensing. In addition, the few-layer BP (FL-BP) nanosheet constructed field-effect transistor (FET) is exfoliated on Si-based wafer organized with a huge on/off proportion showed the hole mobility (300–1000 cm^2/Vs), and when a BP nanosheet is positioned on a simple hexagonal boron nitride (h-BN)-based target material or sandwiched between two layers structure of h-BN flakes at low temperature, the mobility value is gradually improved and enhanced to 2000–4000 cm^2/Vs, which is extremely near the hole mobility value ($\approx$6500 cm^2/Vs) of bulk BP. These properties and characteristics distinguish the BP nanomaterial from other 2D layered structural materials such as TMDs, BN, and graphene. Consequently, due to these multidirectional properties, BP can be used as an alternative material for minimizing the hurdles during the fabrication process for channels based on large-area membranes in the silicon semiconductor industry (Akahama, Endo, and Narita 1983; Doganov et al. 2015). However, the major bottleneck for more widespread applications of 2D materials is the difficulty of fabricating large-scale structures/devices based on 2D materials because of the lack of major developments in large-scale synthesis of 2D materials. Recently, scientists have started to look beyond these fields, and new research directions of "2D material heterostructures" and "three-dimensional (3D) architecture of 2D-based nanomaterial" have started to emerge. One alternative, to utilize 2D materials as a portion of practical applications, macroscopic devices on a large scale, is required to control the self-assembly of these 2D nanomaterial building blocks into more sophisticated and hierarchical 3D architectures. This competence can be crucial to fabricate the complex structures network, multifunctional devices with tailored physical and chemical properties. However, the building of 3D hierarchical assemblies of the 2D nanosheets is a difficult target. The creation of 3D hierarchical structures and the inter-plane coupling of 2D materials might result in misplacing their distinctive assets. Consequently, when making 3D macro and hybrid structures from 2D materials, a significant and fundamental task is to stop the interlayer restacking in order to preserve their distinctive 2D characteristics. Once stacked in 3D form, 2D materials maintain their intrinsic properties at the micro/nanoscale and invoke new properties due to their macro and hybrid structures (Sun, Xu, and Gao 2013; Xu et al. 2010; Jiang and Fan 2014; Li and Shi 2012). To know about such properties of BP, a lot of research work and significant development has done during the past few years toward understanding the major practical exploration of 2D BP nanomaterials. Due to their previously mentioned multidirectional properties, BP can be established as an alternative material for minimizing the hurdles for designing ultrathin large-area membranes for channels in the silicon semiconductor industry to prepare 3D hybrid structures (Akahama, Endo, and Narita 1983; Doganov et al. 2015). The BP laminated layer/membrane-based 3D hybrid structures has a great van der Waals (vdWs) interaction for other 2D layered materials, quantum dots, polymers, and small organic molecules, and acts as a good host/hybrid structure with enhanced electrical, mechanical, biodegradable properties. Besides, improvement in synthetic protocol and assembly to 3D macrostructures of BP for potential technological applications is still a work in progress. It is further analyzed that the 3D macroscopic structures of 2D materials are porous and can be used as electrodes for batteries/supercapacitor applications (Chen 2013; Wang et al. 2010; Guo, Song, and Chen 2009; Hou et al. 2011; Simon and Gogotsi 2008; Wang, Zhang, and Zhang 2012; Zhang and Zhao 2009). Moreover, the pores can host various other materials, including metals, organic molecules, polymers, etc. (Dong et al. 2012; Yong et al. 2012), resulting in new functionalities and applications. 3D macrostructures of BP will contain the fundamental characteristics of BP with high carrier mobility, narrow bandgap, and mechanical strength, and contain new functionalities due to their porous structure. These pores alleviate the transport and ion diffusion, and the interconnecting 2D nanosheets in the 3D architecture provide abundant paths for electron transport. These attributes are essential for efficient energy conversion and storage materials. So, from the above discussion, we can conclude that while 2D nanosheets of BP materials may have applications in electronic devices for energy conversion/storage and other biomedical applications where large porous scale structures are required, 3D macrostructures of BP can come in handy (Lv et al. 2018; Wen et al. 2019; Yang et al. 2018; Xing et al. 2018). During recent decades, a lot of work has been done toward improving the synthetic protocol and application of low-dimensional materials, including 1D materials (nanotubes, nanowires) (Iijima 1991; Ebbesen and Ajayan 1992; Sandler et al. 1999; Coleman et al. 2006; Spitalsky et al. 2010; Abbasi et al. 2015, 2016) and 0D materials (quantum dots) (Li et al. 2015; Chan and Nie 1998) for potential applications and biomedicine. These materials have received considerable attention among scientific research communities, but scalable production and batch-to-batch reproducibility of these materials remain big problems and challenges. Due to these hurdles, the synthesis of this material from academic studies to industrial applications is still in nascent stages and commercializing nanomaterial-based products is problematic. However, the challenge related to constructing 3D hierarchical assemblies and hybrid structures of the 2D BP is not as easy as the synthesis of comparatively hierarchical structures materials of 2D BP. During the construction of 3D structures, there is the possibility of damaging the original unique properties of BP as a result of inter-plane coupling of 2D BP layered materials (Lei et al. 2017; Eswaraiah et al. 2016; Wu, Hui, and Hui 2018).

Here in this chapter, we summarize the synthesis, structure, and properties of 2D nanomaterial and 3D architecture of BP. We have drawn on the constructive attempt made around the world by different scientific communities and research groups working in nanosciences, which has resulted in a significant number of review articles on the preparation of BP nanomaterials being published. The published data is accomplished in producing a broad range of 2D BP nanomaterial in different morphology, such as quantum dots, nanoflakes, nanosheets, and nanoribbons. This book emphasizes the current developments in the field associated with the syntheses, properties, applications of well-defined 2D and 3D structures of BP. We review in detail the currently available literature of the various synthetic strategies for the 2D BP nanomaterial, their scalable synthesis, and assemblies into 3D macro and hybrid structures. Moreover, the different methodologies for the adornment of the 2D nanostructures of BP with other materials like polymers, graphene transition metal dichalcogenides are discussed, along with a comparative study of the 3D architecture of BP with other 3D layered materials and their applications in various fields.

1.2   OVERVIEW OF CHAPTERS

In order to cover synthesis, physical properties, and applications, we tried to assemble a large sample of semiconducting black phosphorus (BP) research. We begin from basic chemistry of phosphorus; that is, various allotropes of phosphorus, synthesis and characteristics of 2D BP, their heterostructures and complex 3D architecture. The general overview of the chapters is as follows.

In Chapter 2, the authors Nasir Mahmood Abbasi, Han Zhang and Bing Wang summarize the rich structural chemistry of many existing allotropes and the complex geometry of coordination. It is mentioned that white, black, and red are the three primary allotropes of phosphorus, and the toxic and corrosive characteristic of white phosphorus is inappropriate for use as anode material. Black phosphorus has a higher conductivity and is more thermodynamically and chemical stable than red phosphorus. However, its synthesis is difficult and requires high temperatures and strain, which means that the three varieties are commercially of the lowest value. Red phosphorus is chemically stable, abundant, inexpensive and environmentally friendly, as compared to white and black phosphorus, making it a promising candidate for high-energy lithium-ion batteries (LIBs). In spite of those advantages, the main roadblocks for its use in LIBs are two inherent red phosphorus issues, similar to other intermetallic alloys. First, the electrochemical redox reactions are poor in electronic conductivity (1×10^{-10} S cm^{-1}). Second, its significant expansion in volume of up to 300 percent causes similar problems to those found in the more extensively investigated alloy anodes. Due to this function, an infinite number of allotropes appear to be possible. This shows how similar it is to carbon, the highest numbered element known to allotropic substances, as well as three-dimensional bulk materials, in that two-dimensional, layered members, one-dimensional phosphorus nanorods and nanotubes can be formed. So it's no wonder new allotropic forms are expected and synthesized daily. The chemists working in these areas should concentrate on phosphorus, oxygen, moisture, and light reactivity and sensitivity, particularly for nanostructured materials and appliances. The vast majority of research will therefore concentrate on hybrid materials so that phosphorus allotropes in combination with other phases can be protected and accessed under environmental conditions. In addition, in a whole variety of scientific sectors such as material science, engineering and production, phosphorus must be brought into business in future.

In Chapter 3, different synthetic techniques, such as top-down and bottom-up approaches, as well as BP properties, are discussed. Apart from other characteristics such as semiconductive values and chemical and thermal stability, different structures—including quantic dots, nanotubes, bulk and monolayers—possess remarkable optical and biocompatible electrical and mechanical features. The problems of stability of BP, synthetic policies (including control morphology), and high rates are highly desired to be resolved in the case of practical applications of BP-based nanostructures for use in electronics, photonics, biomedicine, sensing, and environmental applications.

In Chapter 4, the authors summarize the BP-based heterostructures demonstrated successfully, and concluded that there is a need for significant efforts for further optimization of performance. However, to achieve realistic implementations, the same importance should be given to system reliability and longevity. The thorough preparation of stable BP nanosheets (BPNSs) continues to be a challenge, reducing the creation of heterostructures for use. As heterostructures based on BPNSs evolve, it is very important that new methods of synthetic design are used to maintain and manage the properties of each material. BP study is expected to make mixed-dimensional van der Waals (MDWs) a lifelong hotspot for the production of the killer application of BP and lead eventually to major developments in modern nanotechnology and BP-based product marketing. Meanwhile, research into BP-based heterostructures can be a point of reference for other novel 2D crystals beyond BP and may be used for BP-based van der Waals (vdWs) systems as a general performance enhancement technique.

In Chapter 5 the authors summarize that 3D heterostructures are expected to be extensively studied in the coming years in order to develop effective synthetic methods and gain a better understanding of their properties and applications. Developments in BP-based 3D synthesis strategies in recent years are extremely promising, but it is still early. The comparison of BP to other structures in the 3D layer of the material and BP for future and biomedical applications is also coordinated. The lack of large-scale modular self-assembling and scalable device routes on diverse substrates could be addressed through future study.

Chapter 6 provides a brief overview of optoelectronics, energy storage devices, biomedical applications, and catalysis-related applications of BP based heterostructures with polymers, quantum dots, and metal oxides. BP material filled or coated with matrix-like polymers, quantum dots, or metal oxides produces composites and hybrids in the form of 3D structures with enhanced mechanical strength, electrical or thermal conductivity, light absorption or emission, or large-scale manufacturing. The future of BP-based composite materials in photonics technologies is expected to include versatile and transparent substrates with infinite dimensions and long service lives, and lower costs and easier planning (potential and biomedical applications).

In the future, the use of 3D structures based on BP is thus expected to remain a hot trend.

Finally, in Chapter 7 the authors provide a brief summary, including different synthetic procedures (top-down and bottom-up approaches) containing properties of BP: composite of BP with polymers and other 2D layered material, comparison of BP with other 2D layered material, construction of 3D structures of BP, their market demand, drawbacks and applications in the fields of energy and biomedicine are discussed. 2D BP nanomaterials, such as BP nanotubes, bulk and monolayer, and quantum dot BP are discovered to have excellent optical, biocompatible, thermal, mechanical, and electrical properties, as well as nontoxicity, chemical, and thermal stability values. BP compounds with different 3D structures are thus promising candidates for a range of technical and industrial applications, including optoelectronics, energy storage, and biocompatible drug delivery materials.

1.3 SUMMARY

In addition to other properties such as the importance of nontoxicity, chemical and thermal stability, 2D BP nanomaterials like the NP, bulk and monolayer, as well as the quantum dot BP, have outstanding optical, biocompatible, thermal, mechanical, and electrical properties. There are also attractive candidates for various technologies and industrial applications ranging from optoelectronics to energy storage and biocompatible material for the delivery of drugs to the different 3D structures for BP compounds.

1.4 ACKNOWLEDGMENTS

The research was partially supported by financial support from the Science and Technology Development Fund (Nos. 007/2017/A1 and 132/2017/A3), Macao Special Administration Region (SAR), China, and the National Natural Science Fund (Grant Nos. 61875138, 61435010, 61805147, and 6181101252), and Science and Technology Innovation Commission of Shenzhen (KQTD2015032416270, JCYJ20150625103619275, JCYJ20180305125141661, and JCYJ20170811093453105). The authors also acknowledge the support from the Instrumental Analysis Center of Shenzhen University (Xili Campus).

REFERENCES

Abbasi, Nasir M., Haojie Yu, Li Wang, Zain-ul-Abdin, Wael A. Amer, Muhammad Akram, Hamad Khalid, Yongsheng Chen, Muhammad Saleem, Ruoli Sun, and Jie Shan. 2015. "Preparation of silver nanowires and their application in conducting polymer nanocomposites." *Materials Chemistry and Physics* 166: 1–15.

Abbasi, Nasir M., Li Wang, Haojie Yu, Zain-ul-Abdin, Muhammad Akram, Hamad Khalid, Chen Yongshen, Ruoli Sun, Muhammad Saleem, and Zheng Deng. 2016. "Glycerol and water mediated synthesis of silver nanowires in the presence of cobalt chloride as growth promoting additive." *Journal of Inorganic and Organometallic Polymers and Materials* 26(3): 680–690.

Akahama, Yuichi, Shoichi Endo, and Shin-ichiro Narita. 1983. "Electrical properties of black phosphorus single crystals." *Journal of the Physical Society of Japan* 52(6): 2148–2155.

Britnell, L., R. V. Gorbachev, R. Jalil, B. D. Belle, F. Schedin, A. Mishchenko, T. Georgiou, M. I. Katsnelson, L. Eaves, S. V. Morozov, N. M. R. Peres, J. Leist, A. K. Geim, K. S. Novoselov, and L. A. Ponomarenko. 2012. "Field-effect tunneling transistor based on vertical graphene heterostructures." *Science* 335(6071): 947.

Butler, Sheneve Z., Shawna M. Hollen, Linyou Cao, Yi Cui, Jay A. Gupta, Humberto R. Gutiérrez, Tony F. Heinz, Seung Sae Hong, Jiaxing Huang, and Ariel F. Ismach. 2013. "Progress, challenges, and opportunities in two-dimensional materials beyond graphene." *ACS nano* 7(4): 2898–2926.

Castellanos-Gomez, Andres, Leonardo Vicarelli, Elsa Prada, Joshua O. Island, K. L. Narasimha-Acharya, Sofya I. Blanter, Dirk J. Groenendijk, Michele Buscema, Gary A. Steele, and J. V. Alvarez. 2014. "Isolation and characterization of few-layer black phosphorus." *2D Materials* 1(2): 025001.

Chan, Warren C. W., and Shuming Nie. 1998. "Quantum dot bioconjugates for ultrasensitive nonisotopic detection." *Science* 281(5385): 2016–2018.

Chen, Jiajun. 2013. "Recent progress in advanced materials for lithium ion batteries." *Materials* 6(1): 156–183.

Chen, Wansong, Jiang Ouyang, Hong Liu, Min Chen, Ke Zeng, Jianping Sheng, Zhenjun Liu, Yajing Han, Liqiang Wang, and Juan Li. 2017. "Black phosphorus nanosheet-based drug delivery system for synergistic photodynamic/photothermal/chemotherapy of cancer." *Advanced Materials* 29(5): 1603864.

Chen, Yu, Chujun Zhao, Shuqing Chen, Juan Du, Pinghua Tang, Guobao Jiang, Han Zhang, Shuangchun Wen, and Dingyuan Tang. 2013. "Large energy, wavelength widely tunable, topological insulator Q-switched erbium-doped fiber laser." *IEEE Journal of Selected Topics in Quantum Electronics* 20(5): 315–322.

Chen, Yu, Man Wu, Pinghua Tang, Shuqing Chen, Juan Du, Guobao Jiang, Ying Li, Chujun Zhao, Han Zhang, and Shuangchun Wen. 2014. "The formation of various multi-soliton patterns and noise-like pulse in a fiber laser passively mode-locked by a topological insulator based saturable absorber." *Laser Physics Letters* 11(5): 055101.Coleman, Jonathan N., Umar Khan, Werner J. Blau, and Yurii K. Gun'ko. 2006. "Small but strong: a review of the mechanical properties of carbon nanotube–polymer composites." *Carbon* 44(9): 1624–1652.

Doganov, Rostislav A., Eoin C. T. O'Farrell, Steven P. Koenig, Yuting Yeo, Angelo Ziletti, Alexandra Carvalho, David K. Campbell, David F. Coker, Kenji Watanabe, Takashi Taniguchi, Antonio H. Castro Neto, and Barbaros Özyilmaz. 2015. "Transport properties of pristine few-layer black phosphorus by van der Waals passivation in an inert atmosphere." *Nature Communications* 6: 6647.

Dong, X. C., J. X. Wang, J. Wang, M. B. Chan-Park, X. G. Li, L. H. Wang, W. Huang, and P. Chen. 2012. "Supercapacitor electrode based on three-dimensional graphene-polyaniline hybrid." *Materials Chemistry and Physics* 134(2–3): 576–580.

Ebbesen, T. W., and P. M. Ajayan. 1992. "Large-scale synthesis of carbon nanotubes." *Nature* 358(6383): 220–222.

Eswaraiah, Varrla, Qingsheng Zeng, Yi Long, and Zheng Liu. 2016. "Black phosphorus nanosheets: synthesis, characterization and applications." *Small* 12(26): 3480–3502.

Guo, Peng, Huaihe Song, and Xiaohong Chen. 2009. "Electrochemical performance of graphene nanosheets as anode material for lithium-ion batteries." *Electrochemistry Communications* 11(6): 1320–1324.

Guo, Zhinan, Si Chen, Zhongzheng Wang, Zhenyu Yang, Fei Liu, Yanhua Xu, Jiahong Wang, Ya Yi, Han Zhang, and Lei Liao. 2017. "Metal-ion-modified black phosphorus with enhanced stability and transistor performance." *Advanced Materials* 29(42): 1703811.

Hou, Junbo, Yuyan Shao, Michael W. Ellis, Robert B. Moore, and Baolian Yi. 2011. "Graphene-based electrochemical energy conversion and storage: fuel cells, supercapacitors and lithium ion batteries." *Physical Chemistry Chemical Physics* 13(34): 15384–15402.

Iijima, Sumio. 1991. "Helical microtubules of graphitic carbon." *Nature* 354(6348): 56–58.

Jiang, L., and Z. Fan. 2014. "Design of advanced porous graphene materials: from graphene nanomesh to 3D architectures." *Nanoscale* 6(4): 1922–1945.

Kis, Andras. 2012. "Graphene is not alone." *Nature Nanotechnology* 7: 683.

Koenig, Steven P., Rostislav A. Doganov, Hennrik Schmidt, A. H. Castro Neto, and Barbaros Özyilmaz. 2014. "Electric field effect in ultrathin black phosphorus." *Applied Physics Letters* 104(10): 103106.

Kong, Lingchen, Zhipeng Qin, Guoqiang Xie, Zhinan Guo, Han Zhang, Peng Yuan, and Liejia Qian. 2016. "Black phosphorus as broadband saturable absorber for pulsed lasers from 1 μm to 2.7 μm wavelength." *Laser Physics Letters* 13(4): 045801.

Lei, Wanying, Gang Liu, Jin Zhang, and Minghua Liu. 2017. "Black phosphorus nanostructures: recent advances in hybridization, doping and functionalization." *Chemical Society Reviews* 46(12): 3492–3509.

Li, Chun, and Gaoquan Shi. 2012. "Three-dimensional graphene architectures." *Nanoscale* 4(18): 5549–5563.

Li, Jianfeng, Hongyu Luo, Bo Zhai, Rongguo Lu, Zhinan Guo, Han Zhang, and Yong Liu. 2016. "Black phosphorus: a two-dimension saturable absorption material for mid-infrared Q-switched and mode-locked fiber lasers." *Scientific Reports* 6: 30361.

Li, Likai, Yijun Yu, Guo Jun Ye, Qingqin Ge, Xuedong Ou, Hua Wu, Donglai Feng, Xian Hui Chen, and Yuanbo Zhang. 2014. "Black phosphorus field-effect transistors." *Nature Nanotechnology* 9: 372–377.

Li, Likai, Guo Jun Ye, Vy Tran, Ruixiang Fei, Guorui Chen, Huichao Wang, Jian Wang, Kenji Watanabe, Takashi Taniguchi, Li Yang, Xian Hui Chen, and Yuanbo Zhang. 2015. "Quantum oscillations in a two-dimensional electron gas in black phosphorus thin films." *Nature Nanotechnology* 10: 608.

Liu, Han, Adam T. Neal, Zhen Zhu, Zhe Luo, Xianfan Xu, David Tománek, and Peide D. Ye. 2014. "Phosphorene: an unexplored 2D semiconductor with a high hole mobility." *ACS Nano* 8(4): 4033–4041.

Liu, Jun, Yu Chen, Pinghua Tang, Changwen Xu, Chujun Zhao, Han Zhang, and Shuangchun Wen. 2015. "Generation and evolution of mode-locked noise-like square-wave pulses in a large-anomalous-dispersion Er-doped ring fiber laser." *Optics Express* 23(5): 6418–6427.

Liu, Meng, Nian Zhao, Hao Liu, Xu-Wu Zheng, Ai-Ping Luo, Zhi-Chao Luo, Wen-Cheng Xu, Chu-Jun Zhao, Han Zhang, and Shuang-Chun Wen. 2014. "Dual-wavelength harmonically mode-locked fiber laser with topological insulator saturable absorber." *IEEE Photonics Technology Letters* 26(10): 983–986.

Lu, Jiong, Kai Zhang, Xin Feng Liu, Han Zhang, Tze Chien Sum, Antonio H Castro Neto, and Kian Ping Loh. 2013. "Order–disorder transition in a two-dimensional boron–carbon–nitride alloy." *Nature Communications* 4: 2681.

Lv, Zhisheng, Yuxin Tang, Zhiqiang Zhu, Jiaqi Wei, Wenlong Li, Huarong Xia, Ying Jiang, Zhiyuan Liu, Yifei Luo, and Xiang Ge. 2018. "Honeycomb-lantern-inspired 3D stretchable supercapacitors with enhanced specific areal capacitance." *Advanced Materials* 30(50): 1805468.

Ma, Jie, Shunbin Lu, Zhinan Guo, Xiaodong Xu, Han Zhang, Dingyuan Tang, and Dianyuan Fan. 2015. "Few-layer black phosphorus based saturable absorber mirror for pulsed solid-state lasers." *Optics Express* 23(17): 22643–22648.

Mao, Dong, Biqiang Jiang, Xuetao Gan, Chaojie Ma, Yu Chen, Chujun Zhao, Han Zhang, Jianbang Zheng, and Jianlin Zhao. 2015. "Soliton fiber laser mode locked with two types of film-based Bi 2 Te 3 saturable absorbers." *Photonics Research* 3(2): A43–A46.

Mao, Nannan, Jingyi Tang, Liming Xie, Juanxia Wu, Bowen Han, Jingjing Lin, Shibin Deng, Wei Ji, Hua Xu, and Kaihui Liu. 2015. "Optical anisotropy of black phosphorus in the visible regime." *Journal of the American Chemical Society* 138(1): 300–305.

Mu, Haoran, Zhiteng Wang, Jian Yuan, Si Xiao, Caiyun Chen, Yu Chen, Yao Chen, Jingchao Song, Yusheng Wang, and Yunzhou Xue. 2015. "Graphene–Bi2Te3 heterostructure as saturable absorber for short pulse generation." *ACS Photonics* 2(7): 832–841.

Neto, A. H., and K. Novoselov. 2011. "Two-dimensional crystals: beyond graphene." *Materials Express* 1(1): 10–17.

Neto, A. H., Francisco Guinea, Nuno M. R. Peres, Kostya S. Novoselov, and Andre K. Geim. 2009. "The electronic properties of graphene." *Reviews of Modern Physics* 81(1): 109.

Novoselov, Kostya S., D. Jiang, F. Schedin, T. J. Booth, V. V. Khotkevich, S. V. Morozov, and Andre K. Geim. 2005. "Two-dimensional atomic crystals." *Proceedings of the National Academy of Sciences* 102(30): 10451–10453.

Ponraj, Joice Sophia, Zai-Quan Xu, Sathish Chander Dhanabalan, Haoran Mu, Yusheng Wang, Jian Yuan, Pengfei Li, Siddharatha Thakur, Mursal Ashrafi, and Kenneth Mccoubrey. 2016. "Photonics and optoelectronics of two-dimensional materials beyond graphene." *Nanotechnology* 27(46): 462001.

Qiao, Jingsi, Xianghua Kong, Zhi-Xin Hu, Feng Yang, and Wei Ji. 2014. "High-mobility transport anisotropy and linear dichroism in few-layer black phosphorus." *Nature Communications* 5: 4475.

Rahman, Mohammad Ziaur, Chi Wai Kwong, Kenneth Davey, and Shi Zhang Qiao. 2016. "2D phosphorene as a water splitting photocatalyst: fundamentals to applications." *Energy & Environmental Science* 9(3): 709–728.

Ren, Xiaohui, Zhongjun Li, Zongyu Huang, David Sang, Hui Qiao, Xiang Qi, Jianqing Li, Jianxin Zhong, and Han Zhang. 2017. "Environmentally robust black phosphorus nanosheets in solution: application for self-powered photodetector." *Advanced Functional Materials* 27(18): 1606834.

Rudenko, Alexander N., and Mikhail I. Katsnelson. 2014. "Quasiparticle band structure and tight-binding model for single- and bilayer black phosphorus." *Physical Review B* 89(20): 201408.

Sandler, J., M. S. P. Shaffer, T. Prasse, W. Bauhofer, K. Schulte, and A. H. Windle. 1999. "Development of a dispersion process for carbon nanotubes in an epoxy matrix and the resulting electrical properties." *Polymer* 40(21): 5967–5971.

Shao, Jundong, Liping Tong, Siying Tang, Zhinan Guo, Han Zhang, Penghui Li, Huaiyu Wang, Chang Du, and Xue-Feng Yu. 2015. "PLLA nanofibrous paper-based plasmonic substrate with tailored hydrophilicity for focusing SERS detection." *ACS Applied Materials & Interfaces* 7(9): 5391–5399.

Shao, Jundong, Hanhan Xie, Hao Huang, Zhibin Li, Zhengbo Sun, Yanhua Xu, Quanlan Xiao, Xue-Feng Yu, Yuetao Zhao, Han Zhang, Huaiyu Wang, and Paul K. Chu. 2016. "Biodegradable black phosphorus-based nanospheres for in vivo photothermal cancer therapy." *Nature Communications* 7: 12967.

Simon, Patrice, and Yury Gogotsi. 2008. "Materials for electrochemical capacitors." *Nature Materials* 7(11): 845–854.

Song, Yu Feng, Han Zhang, and Ding Yuan Tang. 2012. "Polarization rotation vector solitons in a graphene mode-locked fiber laser." *Optics Express* 20(24): 27283–27289.

Spitalsky, Zdenko, Dimitrios Tasis, Konstantinos Papagelis, and Costas Galiotis. 2010. "Carbon nanotube–polymer composites: chemistry, processing, mechanical and electrical properties." *Progress in Polymer Science* 35(3): 357–401.

Splendiani, Andrea, Liang Sun, Yuanbo Zhang, Tianshu Li, Jonghwan Kim, Chi-Yung Chim, Giulia Galli, and Feng Wang. 2010. "Emerging photoluminescence in monolayer MoS2." *Nano Letters* 10(4): 1271–1275.

Sun, Haiyan, Zhen Xu, and Chao Gao. 2013. "Multifunctional, ultra-flyweight, synergistically assembled carbon aerogels." *Advanced Materials* 25(18): 2554–2560.

Sun, Zhengbo, Hanhan Xie, Siying Tang, Xue-Feng Yu, Zhinan Guo, Jundong Shao, Han Zhang, Hao Huang, Huaiyu Wang, and Paul K. Chu. 2015. "Ultrasmall black phosphorus quantum dots: synthesis and use as photothermal agents." *Angewandte Chemie International Edition* 54(39): 11526–11530.

Wan, Pengbo, Xuemei Wen, Chaozheng Sun, Bevita K. Chandran, Han Zhang, Xiaoming Sun, and Xiaodong Chen. 2015. "Flexible transparent films based on nanocomposite networks of polyaniline and carbon nanotubes for high-performance gas sensing." *Small* 11(40): 5409–5415.

Wang, Baolin, Haohai Yu, Han Zhang, Chujun Zhao, Shuangchun Wen, Huaijin Zhang, and Jiyang Wang. 2014. "Topological insulator simultaneously Q-switched dual-wavelength." *IEEE Photonics Journal* 6(3): 1–7.

Wang, Guoping, Lei Zhang, and Jiujun Zhang. 2012. "A review of electrode materials for electrochemical supercapacitors." *Chemical Society Reviews* 41(2): 797–828.

Wang, Hailiang, Li-Feng Cui, Yuan Yang, Hernan Sanchez Casalongue, Joshua Tucker Robinson, Yongye Liang, Yi Cui, and Hongjie Dai. 2010. "Mn3O4– graphene hybrid as a high-capacity anode material for lithium ion batteries." *Journal of the American Chemical Society* 132(40): 13978–13980.

Wang, Qing Hua, Kourosh Kalantar-Zadeh, Andras Kis, Jonathan N. Coleman, and Michael S. Strano. 2012. "Electronics and optoelectronics of two-dimensional transition metal dichalcogenides." *Nature Nanotechnology* 7: 699.

Wang, Qingkai, Yu Chen, Lili Miao, Guobao Jiang, Shuqing Chen, Jun Liu, Xiquan Fu, Chujun Zhao, and Han Zhang. 2015. "Wide spectral and wavelength-tunable dissipative soliton fiber laser with topological insulator nano-sheets self-assembly films sandwiched by PMMA polymer." *Optics Express* 23(6): 7681–7693.

Wang, Renheng, Xinhai Li, Zhixing Wang, and Han Zhang. 2017. "Electrochemical analysis graphite/electrolyte interface in lithium-ion batteries: p-Toluenesulfonyl isocyanate as electrolyte additive." *Nano Energy* 34: 131–140.

Wang, X., A. M. Jones, K. L. Seyler, V. Tran, Y. Jia, H. Zhao, H. Wang, L. Yang, X. Xu, and F. Xia. 2015. "Highly anisotropic and robust excitons in monolayer black phosphorus." *Nature Nanotechnoly* 10(6): 517–521.

Wang, Z. T., Y. Chen, C. J. Zhao, H. Zhang, and S. C. Wen. 2012. "Switchable dual-wavelength synchronously Q-switched erbium-doped fiber laser based on graphene saturable absorber." *IEEE Photonics Journal* 4(3): 869–876.

Wen, Min, Danni Liu, Yihong Kang, Jiahong Wang, Hao Huang, Jia Li, Paul K. Chu, and Xue-Feng Yu. 2019. "Synthesis of high-quality black phosphorus sponges for all-solid-state supercapacitors." *Materials Horizons* 6(1): 176–181.

Wu, Shuxing, Kwan San Hui, and Kwun Nam Hui. 2018. "2D black phosphorus: from preparation to applications for electrochemical energy storage." *Advanced Science* 5(5): 1700491.

Xia, Fengnian, Han Wang, and Yichen Jia. 2014. "Rediscovering black phosphorus as an aniso-tropic layered material for optoelectronics and electronics." *Nature Communications* 5: 4458.

Xiang, Yuanjiang, Xiaoyu Dai, Jun Guo, Han Zhang, Shuangchun Wen, and Dingyuan Tang. 2014. "Critical coupling with graphene-based hyperbolic metamaterials." *Scientific Reports* 4: 5483.

Xing, Chenyang, Shiyou Chen, Meng Qiu, Xin Liang, Quan Liu, Qingshuang Zou, Zhongjun Li, Zhongjian Xie, Dou Wang, and Biqin Dong. 2018. "Conceptually novel black phosphorus/cellulose hydrogels as promising photothermal agents for effective cancer therapy." *Advanced Healthcare Materials* 7(7): 1701510.

Xu, Mingsheng, Tao Liang, Minmin Shi, and Hongzheng Chen. 2013. "Graphene-like two-dimensional materials." *Chemical Reviews* 113(5): 3766–3798.

Xu, Yuxi, Kaixuan Sheng, Chun Li, and Gaoquan Shi. 2010. "Self-assembled graphene hydrogel via a one-step hydrothermal process." *ACS Nano* 4(7): 4324–4330.

Yan, Peiguang, Rongyong Lin, Hao Chen, Han Zhang, Aijiang Liu, Haipeng Yang, and Shuangchen Ruan. 2014. "Topological insulator solution filled in photonic crystal fiber for passive mode-locked fiber laser." *IEEE Photonics Technology Letters* 27(3): 264–267.

Yang, Bowen, Junhui Yin, Yu Chen, Shanshan Pan, Heliang Yao, Youshui Gao, and Jianlin Shi. 2018. "2D-black-phosphorus-reinforced 3D-printed scaffolds: a stepwise counter-measure for osteosarcoma." *Advanced Materials* 30(10): 1705611.

Yong, Yang-Chun, Xiao-Chen Dong, Mary B. Chan-Park, Hao Song, and Peng Chen. 2012. "Macroporous and monolithic anode based on polyaniline hybridized three-dimensional graphene for high-performance microbial fuel cells." *ACS Nano* 6(3): 2394–2400.

Zhang, H., D. Y. Tang, L. M. Zhao, and H. Y. Tam. 2008. "Induced solitons formed by cross-polarization coupling in a birefringent cavity fiber laser." *Optics Letters* 33(20): 2317–2319.

Zhang, Han, Ding Yuan Tang, L. M. Zhao, and Xuan Wu. 2009. "Observation of polariza-tion domain wall solitons in weakly birefringent cavity fiber lasers." *Physical Review B* 80(5): 052302.

Zhang, Han, Ding Yuan Tang, Lu Ming Zhao, and Xuan Wu. 2011. "Dual-wavelength domain wall solitons in a fiber ring laser." *Optics Express* 19(4): 3525–3530.

Zhang, Li, and X. S. Zhao. 2009. "Carbon-based materials as supercapacitor electrodes." *Chemical Society Reviews* 38(9): 2520–2531.

Zhao, L. M., D. Y. Tang, Han Zhang, T. H. Cheng, H. Y. Tam, and Chao Lu. 2007. "Dynamics of gain-guided solitons in an all-normal-dispersion fiber laser." *Optics Letters* 32(13): 1806–1808.

Zhao, L. M., D. Y. Tang, H. Zhang, and X. Wu. 2008a. "Polarization rotation locking of vector solitons in a fiber ring laser." *Optics Express* 16(14): 10053–10058.

Zhao, L. M., D. Y. Tang, H. Zhang, X. Wu, and N. Xiang. 2008b. "Soliton trapping in fiber lasers." *Optics Express* 16(13): 9528–9533.

Zhao, L. M., D. Y. Tang, X. Wu, Han Zhang, and H. Y. Tam. 2009a. "Coexistence of polarization-locked and polarization-rotating vector solitons in a fiber laser with SESAM." *Optics Letters* 34(20): 3059–3061.Zhao, L. M., D. Y. Tang, H. Zhang, and X. Wu. 2009b. "Bunch of restless vector solitons in a fiber laser with SESAM." *Optics Express* 17(10): 8103–8108.

Zhao, L. M., D. Y. Tang, X. Wu, and H. Zhang. 2010. "Dissipative soliton generation in Yb-fiber laser with an invisible intracavity bandpass filter." *Optics Letters* 35(16): 2756–2758.

2 Chemistry of Phosphorus

2.1 INTRODUCTION

In industrial applications and processes, phosphorus-containing compounds are commonly used. Hexafluoridophosphates are used as fertilizers and food additives, and hexafluoridophosphates are used as electrolytes in batteries, to name a few applications (Nilges, Schmidt, and Weihrich 2011). Phosphorus (P) is one of the periodic table's strangest elements. The periodic table's 15th element, it was first distilled during a scientist's search for gold. To date, phosphorus is the chemical element with the most unstable thermodynamic allotrope, whereas the other modifications have negative standard enthalpy values. Hennig Brand removed white phosphorus from urine in 1669, which began the history of phosphorus. It was a groundbreaking discovery at the time—imagine a solid that glows for no apparent cause.

The evolution of phosphorus (P) structural chemistry dates back to the nineteenth century. In 1848, amorphous red phosphorus was discovered, which was made by heating white phosphorus. A few years later, Hittorf described and illustrated another kind of phosphorus (Hittorf 1865), which we now call "violet" or "Hittorf phosphorus." Bridgman first mentioned black phosphorus in 1914, which is the most dense and thermodynamically stable form of the element (Bridgman 1914). During the second half of the twentieth century and the early twenty-first century, more allotropes were discovered experimentally (Nilges, Schmidt, and Weihrich 2011).

Phosphorus shows fascinating structural diversity and attention continues to be drawn to the discovery of new changes. Phosphorus is a non-metal that comes in a variety of allotropic forms, from highly reactive white phosphorus to thermodynamically stable black phosphorus (BP).The phosphorus element fascinates with a wide range of allotropic alterations in layers in BP (Brown and Rundqvist 1965; Jamieson 1963), white tetrahedral phosphorus (α-, β-, and β-P4) (Simon, Borrmann, and Craubner 1987; Simon, Borrmann, and Horakh 1997), or tubular layers in red phosphorus. Some of the red phases remain unknown in structure. Recently, paired tubes with a parallel orientation were identified and obtained in crystalline form (the fibrous P [10] of Ruck), as opposed to the perpendicular arrangement of the violet phosphorus of Hittorf (Ruck et al. 2005). The discovery of different types of phosphorus (P) nanorods (Möller and Jeitschko 1986; Pfitzner and Freudenthaler 1995)

DOI: 10.1201/9781003217145-2

has extended the diversity of P structures but left some unanswered questions about their crystal structures and their place in the P allotrope stability order.

A benchmark case for the performance of standard density-functional theory (DFT) levels with strong covalent networking connected by van der Waals forces in low-dimensional systems serves the phosphorus allotropes. This chapter presents the correct stability description and prediction of the toxicity of discovered P nanorod allotropes (Bachhuber et al. 2014a) for allotropical solid structures. The detailed about various allotrope of phosphorus is mentioned in the following sections.

2.2 ALLOTROPES OF PHOSPHORUS

A key participant in group V of the periodic table, phosphorus (P) accounts for approximately 0.1 percent of the earth's crust. In particular, P occurs in three major allotropic forms, i.e., red phosphorus (RP), white phosphorus (WP), and black phosphorus (BP), as well as violet phosphorus, and A7. To date, different phosphorus allotropes are known and the name of the various forms is rather confusing and unsystematic. There are three main ways of illustrating and distinguishing known forms of phosphorus present in literature and used in principle. The colors of the material, finished in white, red, violet, or black phosphorus, are usually the most common and useable notation. A second method uses consecutive numbers that are based on or related to a certain property, such as thermal stability and the trend toward phase transformation. Finally, the Greek letter notation is also used in a given set or group of structurally linked polymorphs. Various allotropic forms of phosphorus and their crystal structures along with characteristic structural fragments are shown in Figure 2.1a and 2.1b, and their detail is mentioned in the following sections (Sofer et al. 2016; Chowdhury and Datta 2017; Piro et al. 2006).

2.2.1 WHITE PHOSPHORUS

White phosphorus (WP) exists as three modifications at different temperatures and pressures. Among various allotropes of phosphorus, P_4 molecules, i.e. α-P_4, is the most reactive and plastic crystal with the P_4 molecules dynamically rotating around their centers of gravity and usually obtained as a result of the room temperature modification (Nitschke 2010). These molecules having a similar arrangement as in the case of the α–Mn structure (von Schnering 1981). Another low-temperature modification is called β–P_4 Moreover, below 197 K at ambient temperature or higher than 1.0 GPa at ambient pressure, this modification is light stable (Bridgman 1914). Besides, the molecules have fixed orientations in the crystal structure of β–P_4 (space group P11) (Spiess, Grosescu, and Haeberlen 1974). Spiess et al. reported the existence of another low-temperature modification called γ-P_4 (Spiess, Grosescu, and Haeberlen 1974).

The phase relationships among these three modifications were established under ambient pressure by X-ray powder diffraction and DTA experiments. On slow warming, the a-modification soon transforms to γ-P_4. Consequently, up to approximately 160 K, this g-modification exists as a stable lower temperature modification. The phase transformation process is taking place, and it was noticed that at

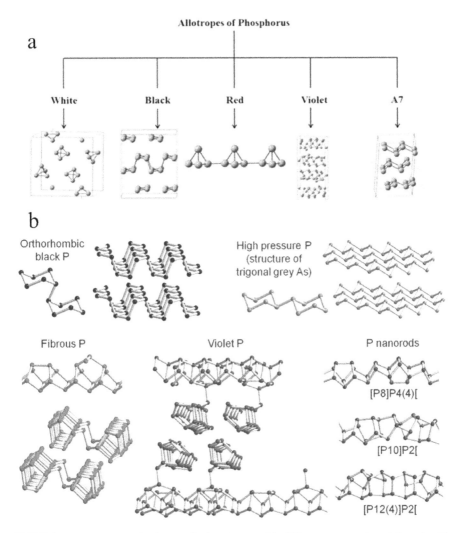

FIGURE 2.1 a) Allotropic forms of phosphorus with different geometries. Adapted with permission from Chowdhury and Datta (2017), Copyright © 2017, American Chemical Society. b) Fragmentation pattern of different P allotropes. Reprinted with permission from (Bachhuber et al. 2014a).

higher temperatures, γ-P_4 irreversibly transforms to β–P_4, and finally β–P_4 reversibly transforms back to α–P_4 at a temperature of about 193 K (Okudera, Dinnebier, and Simon 2005).

2.2.1.1 Toxicity

The toxicity of white phosphorus is similar to cyanide, with a long-term familiarity that might be responsible for causing dreadful bone necrosis (Nitschke 2010). Due to its poison and chemically unstable nature, white phosphorus can't be used as anode material. Moreover, it was reported that after the inhalation process of contents/traces of white phosphorus smoke (185–592 mg m¢3) for between 5 to 15 minutes could cause serious problems including headache, nasal discharge, and upper respiratory tract irritation (Latiff et al. 2015).

2.2.2   RED PHOSPHORUS

Unlike various white and black categories, red phosphorus (RP) is chemically stable, abundant, eco-friendly, and low-cost. Various oxidizing agents react with red phosphorus (RP) vigorously depending on the strength of oxidizing agent. Consequently, in most cases, the product of the reaction causes an instant explosion or fire. Second, the resultant reaction product blasts as a result of interaction with some kind of mechanical shock. Additionally, RP's combustion process produces smoke (tetraphosphorus decaoxide) as the final product, which is corrosive to the skin, respiratory tract, and eyes while found to be non-toxic (Xia et al. 2015; Young 2004).

2.2.3   BLACK PHOSPHORUS

BP differentiates from other 2D layered counterparts such as graphene, silicene, and germanene due to its puckered structure and three crystalline shape, such as simple cubic, rhombohedral, and orthorhombic. The puckered geometry, like construction, along with the armchair (AM) track and local bonding configurations, are responsible for this structural anisotropy. In BP structure, two adjacent inter-plane P atoms are covalently linked with each P atom and on the other side another P atom in an adjacent inter-plane using its p-orbitals. In contrast, weak van der Waals (vdWs) forces are responsible for stacking individual BP layers (Figure 2.2a). Consequently, the sp^3 orbital hybridization is responsible for a puckered honeycomb structure of BP, with zigzag and armchair arrangements beside both the x- and y-axial tracks, respectively (Rahman et al. 2016; Mao et al. 2016). Along the z-direction, the top view of BP exhibits a hexagonal structure having bond angles of 96.3° and 102.1° (Figure 2.2b). The single-layer BP includes two kinds of P–P bonds and two atomic layers. The bond length between two end-to-end P–P is 2.22 Å. The bond length (0.2244 nm)

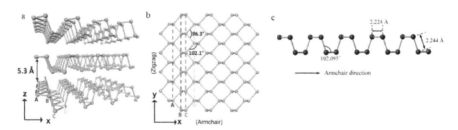

FIGURE 2.2   Structural representation of BP. a) Atomic size structure represents multilayer BP. b) Top view. c) Monolayer phosphorene from a side view. Adapted with permission from Du et al. (2015).

connects the P atoms of top and bottom in a single-layer longer. In comparison, shorter bond length (0.2244 nm) connects the nearest the P atoms in similar plane (Ling et al. 2015). However, these bond lengths are shorter than the relevant dihedral angles (102.095°) alongside the armchair direction and joining bond length (2.24 Å) (Figure 2.2c) with different lattice constants alongside two perpendicular directions at 3.30 Å and 4.53 Å, respectively. The unique anisotropic physical properties of BP are due to this unique structural arrangement (Eswaraiah et al. 2016; Fei and Yang 2014; Rodin, Carvalho, and Castro Neto 2014; Qin et al. 2014; Jiang and Park 2014; Kou, Chen, and Smith 2015). And it was found that BP is the most thermodynamically and chemically stable, and the value of conductivity is higher than the allotrope known as red phosphorus. Although the synthesis of BP needs high pressure and temperature, some easily approachable synthetic route needs to be developed for controlled synthesis of black phosphorus quantum dots (BPQDs), nanotubes, nanoribbons, etc. (Clark and Zaug 2010; Ling et al. 2015).

2.3 STABILITY STUDY OF VARIOUS ALLOTROPES OF PHOSPHORUS

What are the least and most stable phosphorus allotropes? What about the newly discovered and partially unknown forms? Every student will be taught the stability order: white < red < purple < black phosphorus from the published literature. As with many other questions about phosphorus allotropes (Bachhuber et al. 2014b; Aykol, Doak, and Wolverton 2017), this is not certain, however: i) values between 17 and 30 kJ mol^{-1} were reported on the energy difference of white and red phosphorus; (ii) thermodynamics had been identified with five red phosphorus forms, but only structures of two of them were known; the issue of black phosphorus or red phosphorus in all forms, particularly fibrous phosphorus or any other predictable or unknown structure, is still to be found; (iii) new allotropics, such as fibrous and tubular phosphorus, are theoretically expected or predicted; (iv) even recent experimental and theoretical studies do not provide the answers (Aykol, Doak, and Wolverton 2017; Eckstein et al. 2013).

Bridgman (1914), O'Hare (O'Hare and Hubbard 1966; O'Hare, Lewis, and Shirotani 1988) and Jacobs (1937) based their early conclusions on black phosphorus as a stable atmospheric allotrope and at temperature for vapor pressure, calorimetry for combustion and heat formation measurements by reactions with bromine (Jacobs 1937). Ruck's phosphorus is significantly more stable than BP at 400–500° Celsius, according to new research (see the following section) (Figure 2.3a) (Eckstein et al. 2013)). These findings required a response to the questions of whether BP is the most stable allotrope and where the newly discovered allotropes should be listed in the energy spectrum. Phosphorus allotropes, on the other hand, are demanding. The units were combined by Bocker and Haser into molecules (0D), polymers (1D), and layers (2D) (2D units). They discovered that black and violet phosphorus have similar stability, and they predicted fibrous phosphorus and nanorods, which were later discovered by Ruck and Pfitzner. For solid types, modern DFT calculations should be able to predict the correct order of stability to a high degree of accuracy. Pure DFT functionals, on the other hand, fail at first glance. The arsenic-type trigonal

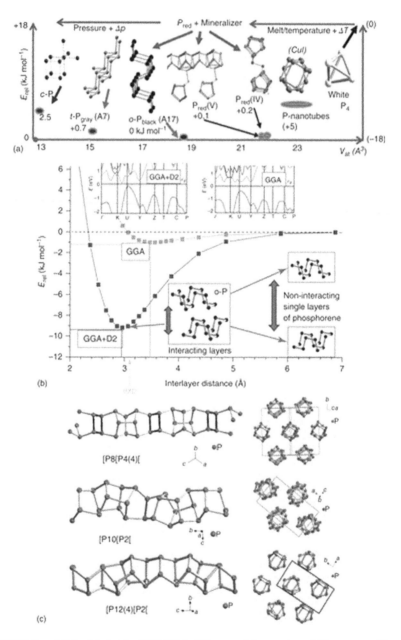

FIGURE 2.3 a) Calculated at the DFT-GGA+D2 stage of theory, the stability ranges of known solid P allotropes. Adapted with permission from Bachhuber et al. (2014) and Nilges, Schmidt, and Weihrich (2011). b) LDA, GGA, and GGA+D2 calculations were used to determine the interlayer and exfoliation capacity. Adapted with permission from Bachhuber et al. (2014), Nilges, Schmidt, and Weihrich (2011), and corresponding structures of electronic band. c) Predicted tubular P crystal structures from simulated CuI extraction. Adapted with permission from Nilges, Schmidt, and Weihrich (2011).

metastable high-pressure form of phosphorus is predicted to be more stable than black phosphorus using local-density approximation (LDA) functionals (Bachhuber, et al. 2014b).

Every tubular form (fibrous, purple, and nanorod) is much more stable than black phosphorus. Obviously, both results contradict experimental observations. The presence of some drawback is the question. They were phosphorus free electron pairs and the corresponding dispersal interactions, which were already decisive when pressurized transitions were made from black phosphorus to trigonal phosphorus (Boulfelfel et al. 2012).

In fact, the treatment of van der Waals interactions, besides covalent interactions, was identified as the key point for determining the stability range for all known solid allotropes of phosphorus (Figure 2.3b (Bachhuber et al. 2014a)). The molecular units, chain or layers attract interactions between van der Waals. Pure DFT theory cannot account for them. In fact, the addition of empirical dispersion corrections by Grimme enabled the comparison with black, violet and red phosphorus to be a reasonable picture. At the theory level of the general approximation of gradients (GGA)+D2, the most stable black phosphorus on the potential energy landscape is found to be = 0 K, which is preferable to the other forms only by 0.1–0.3 kJ mol^{-1} (Bachhuber et al. 2014a).

On this stage of theory, solid white forms of phosphorus are found to be less stable than black P, with an energy gap of 18 kJ mol^{1}. Models for the crystal structures of Pfitzner's nanorod forms may be presented for the first time using molecular modeling of CuI extraction. As a consequence, the rods are arranged in three hexagonal packing schemes (Figure 2.3c). Four-, five-, and six-fold rings are found on single rods, forming eight-, ten-, and 12-fold cages. Haser designated them with the Baudler nomenclature [P8]P4(4)[, [P10]P2, and [P12(4)]P2[, respectively. The solid forms of these novel allotropes are found to be less preferred to black phosphorus by approximately 5 kJ mol^{-1}, contrary to molecular calculations. To the best of our knowledge, this is still the only study that used high-level DFT calculations to include all experimentally identified crystalline allotropes of phosphorus (Bachhuber et al. 2014b).

If the energy van der Waals between the black P layers is changed to GGA+D2, the exfoliation energy to single-layer phosphorene might be calculated. The result was 72 meV/atom on the GGA+D2 level. In fact, the exact value and stability of this dispersion energy remain an issue for research, a cornerstone of scatter-related calculations (Nilges, Schmidt, and Weihrich 2011).

2.4 SUMMARY

Elemental phosphorus shows the rich structural chemistry of many existing allotropes and a varying and complex geometry of coordination. White, black, and red are the three primary allotropes of phosphorus. White phosphorus is poisonous and corrosive, making it unsuitable for use as anode material. Black phosphorus is more thermodynamically and chemically stable than red phosphorus and has a higher conductivity. Its synthesis, however, is difficult and necessitates high temperatures and pressures, resulting in the lowest commercial value of the three types. Red phosphorus, unlike white and black phosphorus, is chemically stable, plentiful, low-cost, and environmentally friendly, making it a promising candidate for high-energy Lithium-ion batteries (LIBs). Despite these benefits, two inherent problems with red phosphorus, close to other intermetallic alloys, are the key roadblocks to its use in LIBs. To begin with, it has a low electronic conductivity (1×10^{-10} S cm^{-1}), making electrochemical redox reactions difficult. Second, its large volume expansion of up to 300 percent causes problems similar to those of the more thoroughly researched alloy anodes. Because of this feature, it seems possible to have an endless number of allotropes. In addition to three dimensional bulk materials, it shows how close it is to carbon, the element with the highest numbers of known allotropic, that it is possible to form two-dimensional, layered representatives, one-dimensional phosphorus nanorods and nanotubes. So, it is no surprise that new allotropic forms are regularly predicted and synthesized. Phosphorus's reactivity and sensitivity to oxygen, moisture, and light, in particular for nanostructured materials and appliances, must be the focus of chemists working in these areas. The vast majority of studies will therefore put their focus into hybrid material, in order to protect and make phosphorus allotropes accessible under ambient conditions, in conjunction with other phases. In future, phosphorus will have to be brought into business in a whole range of scientific areas, such as materials science, engineering, and manufacturing.

2.5 ACKNOWLEDGMENTS

The research was partially supported by financial support from the Science and Technology Development Fund (Nos. 007/2017/A1 and 132/2017/A3), Macao Special Administration Region (SAR), China, and National Natural Science Fund (Grant Nos. 61875138, 61435010, 61805147, and 6181101252), and Science and Technology Innovation Commission of Shenzhen (KQTD20150324162703, JCYJ20150625103619275, JCYJ20180305125141661, and JCYJ20170811093453105). The authors also acknowledge the support from the Instrumental Analysis Center of Shenzhen University (Xili Campus)

REFERENCES

Aykol, Muratahan, Jeff W. Doak, and C. Wolverton. 2017. "Phosphorus allotropes: stability of black versus red phosphorus re-examined by means of the van der Waals inclusive density functional method." *Physical Review B* 95(21): 214115.

Bachhuber, F., J. von Appen, R. Dronskowski, P. Schmidt, Tom Nilges, A. Pfitzner, and R. Weihrich. 2014a. "Van der Waals interactions in selected allotropes of phosphorus." *Zeitschrift für Kristallographie—Crystalline Materials* 230: 107–115.

Bachhuber, F., J. von Appen, R. Dronskowski, P. Schmidt, T. Nilges, A. Pfitzner, and R. Weihrich. 2014b. "The extended stability range of phosphorus allotropes." *Angewandte Chemie International Edition* 53(43): 11629–11633.

Boulfelfel, Salah Eddine, Gotthard Seifert, Yuri Grin, and Stefano Leoni. 2012. "Squeezing lone pairs: The A 17 to A 7 pressure-induced phase transition in black phosphorus." *Physical Review B* 85(1): 014110.

Bridgman, P. W. 1914. "Two new modifications of phosphorus." *Journal of the American Chemical Society* 36(7): 1344–1363.

Brown, Allan, and Stig Rundqvist. 1965. "Refinement of the crystal structure of black phosphorus." *Acta Crystallographica* 19(4): 684–685.

Chowdhury, Chandra, and Ayan Datta. 2017. "Exotic physics and chemistry of two-dimensional phosphorus: phosphorene." *The Journal of Physical Chemistry Letters* 8(13): 2909–2916.

Clark, S. M., and J. M. Zaug. 2010. "Compressibility of cubic white, orthorhombic black, rhombohedral black, and simple cubic black phosphorus." *Physical Review B* 82(13): 134111.

Du, Haiwei, Xi Lin, Zhemi Xu, and Dewei Chu. 2015. "Recent developments in black phosphorus transistors." *Journal of Materials Chemistry C* 3(34): 8760–8775.

Eckstein, Nadine, Andrea Hohmann, Richard Weihrich, Tom Nilges, and Peer Schmidt. 2013. "Synthesis and phase relations of single-phase fibrous phosphorus." *Zeitschrift für anorganische und allgemeine Chemie* 639(15): 2741–2743.

Eswaraiah, Varrla, Qingsheng Zeng, Yi Long, and Zheng Liu. 2016. "Black phosphorus nanosheets: synthesis, characterization and applications." *Small* 12(26): 3480–3502.

Fei, Ruixiang, and Li Yang. 2014. "Strain-engineering the anisotropic electrical conductance of few-layer black phosphorus." *Nano Letters* 14(5): 2884–2889.

Hittorf, W. 1865. "Zur kenntniss des phosphors." *Annalen der Physik* 202(10): 193–228.

Jacobs, Robert B. 1937. "Phosphorus at high temperatures and pressures." *The Journal of Chemical Physics* 5(12): 945–953.

Jamieson, John C. 1963. "Crystal structures adopted by black phosphorus at high pressures." *Science* 139(3561): 1291–1292.

Jiang, Jin-Wu, and Harold S. Park. 2014. "Negative poisson's ratio in single-layer black phosphorus." *Nature Communications* 5: 4727.

Kou, Liangzhi, Changfeng Chen, and Sean C. Smith. 2015. "Phosphorene: fabrication, properties, and applications." *The Journal of Physical Chemistry Letters* 6(14): 2794–2805.

Latiff, Naziah Mohamad, Wei Zhe Teo, Zdenek Sofer, Adrian C. Fisher, and Martin Pumera. 2015. "The cytotoxicity of layered black phosphorus." *Chemistry—A European Journal* 21(40): 13991–13995.

Ling, Xi, Han Wang, Shengxi Huang, Fengnian Xia, and Mildred S. Dresselhaus. 2015. "The renaissance of black phosphorus." *Proceedings of the National Academy of Sciences* 112(15): 4523–4530.

Mao, Nannan, Jingyi Tang, Liming Xie, Juanxia Wu, Bowen Han, Jingjing Lin, Shibin Deng, Wei Ji, Hua Xu, Kaihui Liu, Lianming Tong, and Jin Zhang. 2016. "Optical anisotropy of black phosphorus in the visible regime." *Journal of the American Chemical Society* 138(1): 300–305.

Möller, M. H., and W. Jeitschko. 1986. "Preparation, properties, and crystal structure of the solid electrolytes Cu2P3I2 and Ag2P3I2." *Journal of Solid State Chemistry* 65(2): 178–189.

Nilges, Tom, Peer Schmidt, and Richard Weihrich. 2011. "Phosphorus: the allotropes, stability, synthesis, and selected applications." *Encyclopedia of Inorganic and Bioinorganic Chemistry*. Wiley, 1–18, doi:10.1002/9781119951438.eibc2643.

Nitschke, Jonathan R. 2010. "The two faces of phosphorus." *Nature Chemistry* 3: 90.

O'Hare, P. A. G., and Ward N. Hubbard. 1966. "Fluorine bomb calorimetry. Part 18.—Standard enthalpy of formation of phosphorus pen afluoride and enthalpies of transition between various forms of phosphorus. Thermodynamic functions of phosphorus pentafluoride between 0 and 1500° K." *Transactions of the Faraday Society* 62: 2709–2715.

O'Hare, P. A. G., Brett M. Lewis, and Ichimin Shirotani. 1988. "Thermodynamic stability of orthorhombic black phosphorus." *Thermochimica Acta* 129(1): 57–62.

Okudera, Hiroki, Robert E. Dinnebier, and Arndt Simon. 2005. "The crystal structure of γ-P4, a low temperature modification of white phosphorus." *Zeitschrift für Kristallographie-Crystalline Materials* 220(2/3): 259–264.

Pfitzner, Arno, and Eva Freudenthaler. 1995. "(CuI) 3P12: a solid containing a new polymer of phosphorus predicted by theory." *Angewandte Chemie International Edition in English* 34(15): 1647–1649.

Piro, Nicholas A., Joshua S. Figueroa, Jessica T. McKellar, and Christopher C. Cummins. 2006. "Triple-bond reactivity of diphosphorus molecules." *Science* 313(5791): 1276.

Qin, Guangzhao, Qing-Bo Yan, Zhenzhen Qin, Sheng-Ying Yue, Hui-Juan Cui, Qing-Rong Zheng, and Gang Su. 2014. "Hinge-like structure induced unusual properties of black phosphorus and new strategies to improve the thermoelectric performance." *Scientific Reports* 4: 6946.

Rahman, Mohammad Ziaur, Chi Wai Kwong, Kenneth Davey, and Shi Zhang Qiao. 2016. "2D phosphorene as a water splitting photocatalyst: fundamentals to applications." *Energy & Environmental Science* 9(3): 709–728.

Rodin, A. S, A. Carvalho, and A. H Castro Neto. 2014. "Strain-induced gap modification in black phosphorus." *Physical Review Letters* 112(17): 176801.

Ruck, Michael, Diana Hoppe, Bernhard Wahl, Paul Simon, Yuekui Wang, and Gotthard Seifert. 2005. "Fibrous red phosphorus." *Angewandte Chemie International Edition* 44(46): 7616–7619.

Simon, Arndt, Horst Borrmann, and Hans Craubner. 1987. "Crystal structure of ordered white phosphorus (β-P)." *Phosphorus and Sulfur and the Related Elements* 30(1–2): 507–510.

Simon, Arndt, Horst Borrmann, and Jörg Horakh. 1997. "On the polymorphism of white phosphorus." *Chemische Berichte* 130(9): 1235–1240.

Sofer, Zdenek, Daniel Bouša, Jan Luxa, Vlastimil Mazanek, and Martin Pumera. 2016. "Few-layer black phosphorus nanoparticles." *Chemical Communications* 52(8): 1563–1566.

Spiess, Hans W., R. Grosescu, and H. Haeberlen. 1974. "Molecular motion studied by NMR powder spectra. II. Experimental results for solid P4 and solid Fe (CO) 5." *Chemical Physics* 6(2): 226–234.

von Schnering, Hans Georg. 1981. "Homoatomic bonding of main group elements." *Angewandte Chemie International Edition in English* 20(1): 33–51.

Xia, Dehua, Zhurui Shen, Guocheng Huang, Wanjun Wang, Jimmy C. Yu, and Po Keung Wong. 2015. "Red phosphorus: an earth-abundant elemental photocatalyst for 'green' bacterial inactivation under visible light." *Environmental Science & Technology* 49(10): 6264–6273.

Young, Jay A. 2004. "Red phosphorus." *Journal of Chemical Education* 81(7): 945.

3 2D Form of Black Phosphorus

3.1 INTRODUCTION

Thanks to the efforts of different research groups from all over the world, there exists a wide range of literature about the synthetic protocols of producing Black phosphorous (BP) nanostructures from a single layer to a few layers. Here, in this chapter the main focus is on the recent advancement for improving synthetic methods and different aspects of the extraction of BP nanostructures. Furthermore, most of the reported method is similar in terms of the morphology of the final product, yield, reaction conditions, and mechanistic approach. It is necessary to mention the detail about the reaction necessities. Furthermore, it is important to explore the facile and robust peeling methodologies for large-area phosphorene in a quick way and without any delay in time (Gusmao, Sofer, and Pumera 2017). Despite a lot of research work especially focused on the synthesis of BP-based derivatives done by the different research groups worldwide, research for improving the synthesis strategies (with controlled number of layers and yield) of phosphorene, properties, and BP derivatives applications is still considered to be at an early stage of development. A lot of challenges remain that can be investigated further in future research. This chapter describes the various synthetic techniques of BP including top-down and bottom-up approaches, as well as defect engineering, physical properties, and some general applications of BP. The details about the various synthetic methods for 2D BP, their physical properties and general applications are mentioned in the following sections.

3.2 SYNTHETIC STRATEGIES/MORPHOLOGICAL ENGINEERING AND CHARACTERIZATIONS OF BLACK PHOSPHORUS

The properties of nanomaterials are dependent on their morphology. The term "morphology" deals with the size and shape of the material. If we want to improve the material properties, it is important to improve the material's synthetic strategies in reproducible and scalable synthetic approaches. In the case of 2D BP, a reliable synthetic approach with uniform size and shape is significant for exploring its potential applications and physiochemical properties. Computational calculations like the computational study *Ab initio* predict that phosphorene (single layer BP) has a direct thickness-dependent bandgap of 1.0 eV, confirmed as a result of luminescence

DOI: 10.1201/9781003217145-3

measurements and that these values are significantly larger than the bandgap exhibited by bulk BP (ca. 0.3 eV) (Liu, Neal, et al. 2014; Rudenko and Katsnelson 2014; Dai and Zeng 2014).

To improve the properties and promising applications of 2D BP, 2D BP needs to be fabricated with new reliable synthetic strategies and their assemblies to 3D structures in a large measurable area. A list of reliable synthesis methods of BP is reported in the literature, such as chemical synthesis, mechanical cleavage, and liquid exfoliation. BP nanosheets with the variant quantity of layers and their sizes have been prepared using liquid exfoliation and mechanical methods and multidirectional applications of the products in photoelectric devices and electronics (Sun, Xie, et al. 2015a). However, due to the lack of suitable synthesis strategies, there is still a shortage of scientific literature regarding biomedical and technological applications. All reported methods are categorized under the heading of top-down and bottom-up approaches. The detailed synthetic approaches are discussed in the following sections.

3.2.1 Non-Epitaxial Growth

Usually, the non-epitaxial approach is used where small particles' formation takes place from their bulk counterpart. In the case of BP, the non-epitaxial method is typically used to break down the adhesive and cohesive forces to disperse the arranged coatings/layer and consequently give mono- as well as few-layer nanosheets (Wu, Hui, and Hui 2018; Geim and Novoselov 2007).

3.2.1.1 Mechanical Cleavage Approach

The cleavage through the mechanical approach is the classical approach to get 2D flakes from layered bulk material via exfoliating. In this respect, in 1990, Seibert et al. used transparent tape to fabricate graphite films from the bulk layered sample (Rudenko and Katsnelson 2014).

3.2.1.1.1 Tap Exfoliation

The scotch-tape-based micro cleavage technique was used on a commercially available bulk crystal of BP to achieve single-layer BP (SL-BP) or few-layer BP (FL-BP). Usually known as phosphorene, this is mentioned in Figure 3.1. For safety purposes, acetone, isopropyl alcohol, and methanol were used to eliminate any traces or small content of scotch tape residue from all the samples after a few-layer containing crystals exfoliation of phosphorene. After that, the clean sample is transferred onto the substrate of Si/SiO$_2$. Also, the solvent residue was removed following a 180°C post-bake process in the scotch-tape-based technique (Chen et al. 2015; Liu, Neal, et al. 2014). The atomic force microscopy (AFM) pictures of an exfoliated SL-BP crystal show the surface morphology and the thickness of the layer, as shown in Figure 3.2a. Single-layer phosphorene was confirmed by measuring the step height at the crystal edge of ~0.85 nm. Figure 3.2b shows the photoluminescence (PL) spectrum of exfoliated SL-BP crystal (SL-BP) detected in the region of visible wavelengths.

In a single-layer phosphorene crystal, a pronounced PL signal centered at 1.45 eV bandgap ~100 meV narrow width, is achieved. In contrast, for 10 nm thick BP flakes,

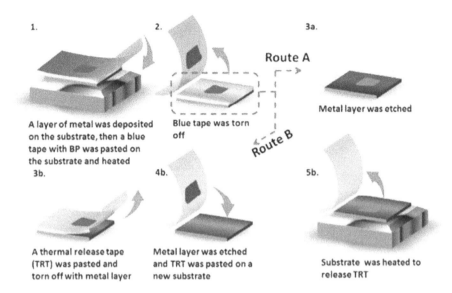

1.
2.
3a.

Route A

A layer of metal was deposited on the substrate, then a blue tape with BP was pasted on the substrate and heated

Blue tape was torn off

Route B

Metal layer was etched

3b.
4b.
5b.

A thermal release tape (TRT) was pasted and torn off with metal layer

Metal layer was etched and TRT was pasted on a new substrate

Substrate was heated to release TRT

FIGURE 3.1 The process of scotch-tape-based micro cleavage technique. Adapted with permission from the Royal Society of Chemistry (Guan et al. 2018).

the expected bandgap of bulk BP is around ~0.3 eV, that value is falling in the wave region of infrared, and there is a deficiency of PL signal within the range of the detected spectrum. This evidence strongly recommended that the PL peak be exciting and on the fundamental bandgap having a minor bound value. The value of bandgap in bulk is meaningfully lower than monolayer and is indirectly confirmed by the measured value of 1.45 eV (Liu, Neal, et al. 2014). Figure 3.2c showed the remaining spectra value of SL-BP, bilayer BP (BL-BP), and bulk BP (B-BP). Furthermore, the spectra show well-defined thickness dependence peak values; with the increase in thickness the peaks such as Ag^1 and Ag^2 have the movable modes value in frequency. Similar results were obtained in the case of MoS_2 (Lee et al. 2010). Moreover, the important concern is how to transfer the mechanically exfoliated materials especially in case of fiber end-facet for all-fiber-based applications. Due to its reliability and simplicity, the mechanical exfoliation process for BP is advantageous because the whole process of fabrication is absent of any costly instruments and chemical procedures.

3.2.1.1.2 Ball Milling

The approach named ball milling (a high-energy consuming mechanical-based milling method) is introduced to synthesis BP from red phosphorus allotrope (Bao et al. 2018). A container having a stainless steel body with 100 stainless steel balls (10 and 5 mm in diameter and 100 g in weight) was introduced for the preparation process. Initially, 0.1 g of red phosphorus was introduced in a container containing polyols (water, ethanol, ethyl alcohol, and ethylene glycol separately) under the Ar atmosphere's influence with a rotation speed of 600 rpm for six hours. The detailed synthesis process is represented in Figure 3.3a. After completing the process, the

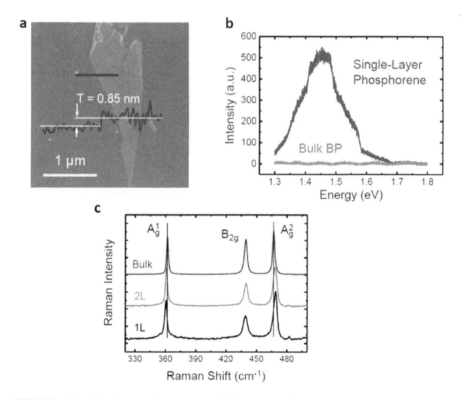

FIGURE 3.2 Few-layer phosphorene (FL-P) and single-layer phosphorene (SL-P) characterizations prepared as a result of tap exfoliation. a) AFM image shows the thickness value of layer ≈0.85 nm of the SL-P crystal. b) Spectra of photoluminescence for bulk BP and SL-P samples after uniformly dispersed on a size around 300 nm mixture of SiO_2/Si glass substrate, represent a pronounced PL signal around 1.45 eV. c) Raman spectra of and bilayer, SL-P, and bulk BP films. Adapted with permission from Liu, Neal, et al. (2014), Copyright © 2014, American Chemical Society.

container was first directly cooled in the ball milling machine for two days and then again in a glove box. After removing the lid, the container was cooled for the same time. The final product in the form of a specific number of samples was analyzed by different techniques such as X-ray diffraction (XRD) to confirm the crystallinity of the final product. Small crystal size, powdered form, red phosphorus having the broadly diffused peak in the final product of BP was confirmed by XRD (Figure 3.3b). After that, as-synthesized black phosphorus quantum dots (BPQDs) with an average lateral size of 6.5 ± 3 nm and a thickness of 3.4 ± 2.6 nm dispersed stably in ethyl alcohol and ethylene glycol was measured from HRTEM (Figure 3.3c–d) and AFM (Figure 3.3e–g) analysis. The d-spacing on HRTEM and electron diffraction pattern match well to the distinguishing crests of BP, such as (111) and (021) (Figure 3.3a). This evidence strongly confirmed that maintaining the uniform size of BP nanocrystals is problematic (Ren et al. 2020).

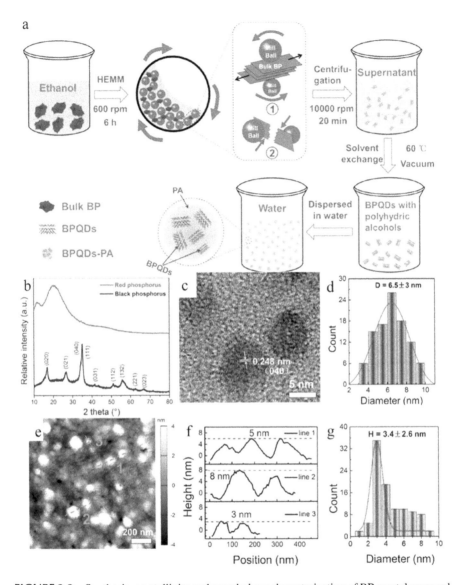

FIGURE 3.3 Synthesis, crystallinity and morphology characterization of BP crystal prepared through ball milling method. a) Schematic representation for the synthesis of BPQDs. b) XRD patterns of BP and amorphous red phosphorus prepared by ball milling method. c) HRTEM images show the internal morphology of large BP crystal and d) represent an average particle size of 6.5 ± 3 nm. e) AFM image of BPQDs. f) Height of BPQDs and g) represent the average thickness. Adapted with permission from Ren et al. (2020), Copyright © 2020, American Chemical Society.

Furthermore, it was noticed that in some parts, BP particle size ranged about 20–50 nm, and the facet of crystals are not clear on the image. It might be due to the process of mechanical milling, the amorphous red phosphorus turned into BP nanocrystals, and the crystals of BP in small size were speedily non-crystallized and extremely spoiled as a result of the electron irradiation process during the transmission electron microscopy (TEM) analysis. Therefore, after 30 minutes of irradiation, only the big grains were left and captured in the TEM image (Ren et al. 2020).

3.2.1.1.3 Wet Chemical Method

This approach synthesizes crystalline nanomaterial mostly in water, without externally added seed crystallites and any surfactant or ligand (Jana, Gearheart, and Murphy 2001; Caswell, Bender, and Murphy 2003; Abbasi et al. 2015). In this method, [Ph$_3$C]BF$_4$ (triphenylcarbenium tetrafluoroborate) and TEMPO (2,2,6,6-tetramethylpiperidine-N-oxyl) were introduced as two electron-deficient organic reagents in solvents such as a combination of water and acetone with equivalent ratio (V: V=1:1) and dichloromethane for BP thinning (high concentration) and passivation (low concentration) in solution. After transferring the mechanically exfoliated BP flakes onto Si/SiO$_2$ substrate, the resultant film was dipped in the solution (containing both [Ph$_3$C]BF4 and TEMPO and in dichloromethane (DCM) and binary solvent system such as a mixture of water and acetone having proper volume ratio (V: V=1:1)). After that, for the chemical modification of BP, two different possible free-radical mechanistic approaches are suggested, shown in Figure 3.4 (Path A and Path B).

In Path A, Ph$_3$C$^+$ breaks neighboring P-P bonds and simultaneously, there is a free radical addition to BP surface by TEMPO molecules. Besides, there is an interaction between the TEMPO molecules and the phosphorus (P) atoms (from BP matrix as the source), creating (Anderson and Shive 1997) the P's positions on the surface of BP. As a result, the process of capturing started, and TEMPO molecules were further captured by showing P electron-deficient species (P radicals) on the surface of BP and so on. Moreover, a hydrolysis type reaction between Ph$_3$C$^+$ and water (using a combination of the binary solvent system (acetone/water (V: V=1:1)) was detected, where the concentration of Ph$_3$C$^+$ can get decreased. As a result, the formation process of the P vacancies is controlled. Conversely, it was found that the increase in the quantity of Ph$_3$C$^+$ controlled the formation of P vacancies when DCM was used as a solvent and, as a result, the breaking of BP surface. On the other side, there is a covalent functionalization process between the oxidized BP surface and TEMPO radical in path B.

Moreover, as a result of the binary solvent system, such as the mixture of water/acetone with equivalent volume ration (V: V=1:1), the expected hydrolysis reaction between water and Ph$_3$C$^+$, the low concentration of Ph$_3$C$^+$ provided a strong pieces of evidence for the controllable oxidation process. Consequently, the TEMPO molecules have posted on the BP surface for the passivation process due to the covalent functionalization process. On the other hand, an oxidation started on the BP surface by the high concentration of [Ph$_3$C]$^+$ when the DCM was used as the solvent; as a result of the oxidation process, there is a deficiency of sharing forces between the

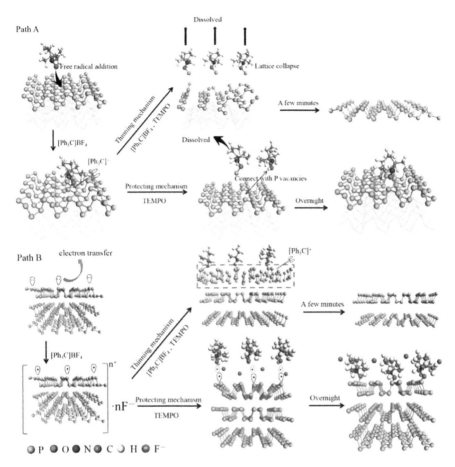

FIGURE 3.4 Schematic representation shows the chemical reaction a mechanistic approach of chemical passivation and thinning process. Adapted with permission from Fan et al. (2019), Copyright © 2019, American Chemical Society.

P–P covalent bonds, and due to that there is a breaking occurring in the structure of BP surface (named as thinning of BP) (Fan et al. 2019).

The AFM analysis of a BP sample shows the surface morphology and variation in thickness of layers after they were previously moved on to the cleaned surface SiO$_2$/Si glass-based substrate (Figure 3.5a) and then (Figure 3.5b) dipped in a solution (a combination of 10 mmol/L [Ph$_3$C]BF$_4$ and 10 mmol/L TEMPO in DCM as a solvent) used for the chemical thinning process.

It was discovered that there is a decrease in the BP-based nanoflake thickness from 10.6 nm to 2.7 nm after chemical thinning (Figure 3.5b). Figure 3.5c–d show that the HR-TEM and the related SAED arrangement confirmed BP's crystallinity due to the chemically thinning process. Before and after the process of chemical thinning, the difference in thickness of the BP samples layers is measured through

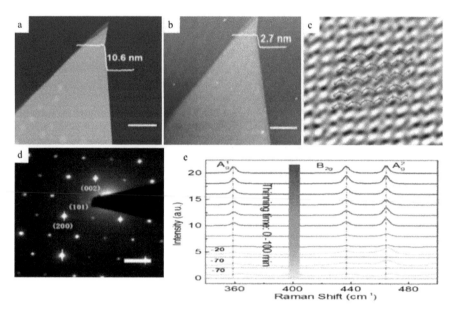

FIGURE 3.5 Morphology analysis of BP crystals as a result of a wet chemical method. AFM images show the surface morphology of a BP nanoflake obtained a) as a result of exfoliation immediately b) after the thinning process for 30 minutes. Both the images having the scale bars range is 1 μm length. c) HR-TEM pictures show the internal morphology of BP crystal with a scale bar of 2 nm. d) SAED images pattern with a scale bars 5 1/nm of the BP nanoflake due to thinning shows the crystalline nature of BP crystal. e) After the different thinning interval, the Raman spectra of a BP nanoflake. Adapted with permission from Fan et al. (2019), Copyright © 2019, American Chemical Society.

AFM image analysis (Figure 3.5f), and it was concluded that the chemical thinning rate was increased gradually as two minutes per layer, which resembles and is comparable to physical thinning methods (Jia et al. 2015). Three signature peaks of the Raman spectra of BP flake are labeled as A_1g, B_2g, and A_2g frequencies and shown in Figure 3.5e, consistent with mechanically exfoliated BP flakes during chemical thinning (Favron et al. 2015). However, it was analyzed that as a result of decreasing the thickness of BP, the cardinal frequency of A_2g slightly redshifts, and it is due to the Davydov splitting on increasing the thinning time to 80, 90, 100 minutes that there is a variation in values of shifting at 0.8, 1.3, and 2.5 cm^{-1} respectively (Favron et al. 2015). To work out the optimum value of the chemical thinning formula, the various combinations of [Ph_3C]BF_4 and TEMPO solute with different concentrations were also tried (Figure 3.6). First, both solutes' concentrations were changed simultaneously from 1 mM to 50 mM in dichloromethane and the color change of BP flakes in the optical images was observed (related to variation in thickness) after two- and seven-hour modifications. It was found that with higher solute concentration, the BP flake was thinned more significantly.

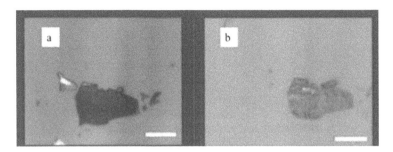

FIGURE 3.6 High concentration of solute effect on the thinning rate. a) Optical microscopy pictures of newly exfoliated BP. b) The same BP flake after treated by 1 M TEMPO and 1 M [Ph3C]BF4 for 3 minutes. Scale bars are 10 μm. The dispersion of water and acetone with equivalent volume ratio (V: V=1:1) (right-most column) for the time duration of 12 hours, respectively. Scale bars: 4 μm 10 Mm TEMPO/10mM[Ph3C]BF$_4$ in H$_2$O acetone (1:1). Adapted with permission from Fan et al. (2019) Copyright © 2019, American Chemical Society.

Second, it was found that on increasing the concentrations of both solute (TEMPO and [Ph$_3$C]BF$_4$ to 1 M in dichloromethane, BP flake was thinned obviously within only three minutes with severely increased surface roughness (Figure 3.6a–b). This evidence strongly suggested that the thinning rate is positively related to the quantity of [Ph$_3$C]BF$_4$. Even after 20 days under the ambient condition of the BP flake being treated by a combination of water and acetone with a fixed value of volume ratio (V:V=1:1) no noticeable degradation was noted. Figure 3.7 shows the AFM images (3.2 nm thick) of a BP flake immediately (Figure 3.7a) and after the chemical passivation of four months (Figure 3.7b). For the chemical passivation process, the solution is a combination of 10 mmol/L [Ph$_3$C]BF$_4$ and 10 mmol/L TEMPO in the binary solvent system, such as the combined suspension of water and acetone with volume combination (V: V=1:1). It is noted that there are no morphology changes observed from the AFM images of passivated BP flake, and it shows a similar value of the surface roughness, such as of 0.66 nm (Figure 3.7a) and 0.68 nm (Figure 3.7b) after extended the time duration for 120 days under the ambient condition.

On the other hand, Figure 3.7c shows the AFM picture of a newly exfoliated bare BP nanoflake (having a thickness value of 5.1 nm) and showing under the ambient condition for 24 hours (Figure 3.7d). Significantly the roughness of the surface gradually increases, ranging from 0.357 nm (Figure 3.7c) to 6.26 nm (Figure 3.7d), and morphology changed was observed. To further support the results, Raman spectroscopy is used to measure the lattice crystalline and stability of the passivated BP under an open atmosphere. There is no variation in intensities values of three functionalized signature peaks of BP flake and stay unaffected even after being treated to air for 120 days (Figure 3.7e) confirmed the air-stability of samples. On the other side, there is variation in intensity values of non-functionalized or bare BP A$_1$g, B$_2$g (Figure 3.7f) decrease considerably after being the treatment with air for just one day, confirmed the morphology change of sample (Fan et al. 2019; Favron et al. 2015).

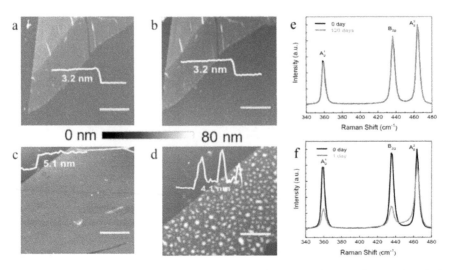

FIGURE 3.7 Surface morphology and thickness of BP nanoflakes. AFM results of a) BP nanoflake as a result of functionalization with no time delay and b) the variations in the morphology of the same sample part of BP nanoflakes unprotected under the ambient environment for the time duration of 120 days (2880h). c) AFM image showing the surface morphology of freshly exfoliated BP flake with no time delay and d) the surface morphology of the same BP flake after one day without any passivation exposed under ambient environment. e) BP nanoflake showed in Raman spectra in (i) (after zero day) for freshly exfoliated BP flake and (ii) (grey line) after passing the 120 days of exfoliation. f) BP nanoflake showed in Raman spectra (i) (after zero day) represents BP flake as a result of fresh exfoliation process (ii) (grey line) after passing the one day of exfoliation. Scale bars are 1 µm. Posted with permission from Fan et al. (2019), Copyright © 2019, American Chemical Society.

3.2.1.2 Liquid Phase Exfoliation

3.2.1.2.1 Sonication Assisted Organic Solvent Exfoliation

The exfoliation process generally consists of three different stages: the dispersal of the material into the solvent, ultrasonication, and purification (Wu, Hui, and Hui 2018). Yasaei et al. (2015) optimized a wide range of solvents having surface tensions (21.7–42.78 dyne cm^{-1}) and polar interaction parameters (2.98–9.3 MPa 1/2) from various chemical functionalities/moieties such as chloro-based organic solvents, alcohols, ketones, cyclic or aliphatic pyrrolidones, and organosulfur compounds, N-alkyl-substituted amides and examined their performance for BP exfoliation. The physical properties of different solvents are illustrated in Table 3.1. A compact mass of BP crystal (0.02 mg/L) was allowed to dip in various organic solvents and treated for sonication for 15 hours with total input/supply energy (around –1 MJ). Thin BP nanoflakes with uniformity and stable dispersion were produced using polar aprotic solvents like dimethyl sulfoxide (DMSO) and dimethylformamide (DMF) after sonication. After that, the supernatants liquid of centrifuged solutions were carefully collected by a syringe pump (Yasaei et al. 2015).

TABLE 3.1
Physical properties of different solvents used for the synthesis of BP nanostructures (Smallwood 2012; Hansen 2007)

Solvents	Surface tension (@20°C dyn/cm)	Boiling point (°C)	Dielectric constant (20°C)	Absolute viscosity (@25°C cP)	Dipole (D)	Solubility in water (25°C %w/w)	Specific heat (cal/mol/°C)	Refractive index (25°C)
Methanol (MeOH)	22.05	64	32.6	0.6	1.7	Total	19.5	1.326
Ethanol (EtOH)	22.18	78	22.4	1.08	1.7	Total	27	1.359
Acetone	22.86	56	20.6	0.33	2.9	Total	30	1.357
2 prapanol (IPA) or isopropyl alcohol	21.7	82	18.3	2.0	1.66	Total	37	1.375
Dimethyl sulfoxide (DMSO)	43.78	189	46.6	2.0	3.96	Total	36	1.476
Dimethylformamide (DMF)	35.20	153	36.7	0.82	3.8	Total	36	1.427
chloroform	27.16	61	4.8	0.57	1.1	0.82	27	1.444
n-hexane	18.4	69	1.9	0.31	0	9.5E-4	42.0	1.372
N-methylpyrrolidone (NMP)	40.7	202	32.2	1.8	4.1	Total	40	1.468

It was further found that in the case of DMSO and DMF, the crystalline product was obtained, while in the case of all other solvents, an amorphous product was found. The amorphous product might be due to the presence of several chemical species in the solvents as mentioned earlier, such as H, F, and Cl, which can strongly bind to the phosphorus atom, acting as scissors to separate "upper" and "lower" (Yasaei et al. 2015). On the other hand, a high boiling point, surface tension, or polar interaction might be responsible for transparent nanoflakes in DMSO and DMF solutions. The physical properties of different solvents used for the exfoliation process of BP are summarized in Table 3.1. Moreover, it was found from the reported literature (Das et al. 2014; Tran et al. 2014; Qiao et al. 2014) that the thickness of layers of BP mainly depends upon direct narrow bandgap, which is ranged from 1 eV in the case of monolayer and around ≈ 0.3 eV in the case of bulk. However, optical absorption spectroscopy was employed to examine the nanoflakes in the range of 830 nm to 2400 nm (near IR) in particularly two solvents (DMF and DMSO).

Interestingly, it was found that many spectral peaks appeared in the NIR range at around ≈ 1.38 eV, ≈ 1.23 eV, ≈ 1.05 eV, ≈ 0.85 eV, and ≈ 0.72 eV for both of the solvents (Figure 3.8a), which were due to the absorption of thick BP nanoflakes of mono to five layers. The detailed study of the layers' thickness distribution was carried out by atomic force microscopy (AFM) (Figure 3.8b–c). In DMSO, flakes of thickness around 15–20 nm were found, while flakes were thinner than 5 nm in DMF (Zhang et al. 2014). It was observed in Raman spectra that the orientation of the sample has a great impact on the intensities of the three vibrational modes. Figure 3.8d represents modes of A1g and A2g after rotating the sample at 180°. By keeping in view the structure of BP, Ag modes were at the maximum level when laser polarization was parallel to the X-axis of crystal (Yasaei et al. 2015).

Moreover, crystallinity, quality, and several layers in BP nanoflakes were analyzed due to TEM images (Figure 3.8e). Furthermore, fast Fourier transform (FFT) was employed to measure the thicknesses of one single layer of BP nanoflakes. As theoretically reported in the literature, the diffraction peaks (110) to (200) having intensity ratio (I 110 / I 200) is smaller than one for multilayer and greater than for monolayer. However, the intensity ratio has only been verified experimentally for multilayers (Castellanos-Gomez et al. 2014). Yasaei et al. (2015) reported the measured ratio of intensities (I 110 / I 200) is around 2.7 ± 0.2, which is quite near to the theoretical value of 2.557 for monolayer BP. The similar features of FFT in various selected areas suggested the existence of BP monolayers, nanoflake, single crystalline, over the whole image (Figure 3.8f) (Castellanos-Gomez et al. 2014).

Kang et al. (2015) determine the optimal solvent for BP exfoliation via tip ultrasonication (Figure 3.9). Seven sets of experiments were conducted using BP crystals in different solvents such as acetone, chloroform, hexane, ethanol, isopropyl alcohol (IPA), dimethylformamide (DMF), and N-methylpyrrolidone (NMP) under ultrasonication.

The effect of the solvent was explored on BP exfoliation by keeping other experimental conditions constant. For safety purposes, moisture-free conditions without O_2 and H_2O were adopted (Figure 3.9a–c). To separate well defined 2D BP nanosheets, the as-prepared resulting dispersions were centrifuged at the conditions of 500–15000 rpm for ten minutes, and consequently, the solution color progressed from brown to

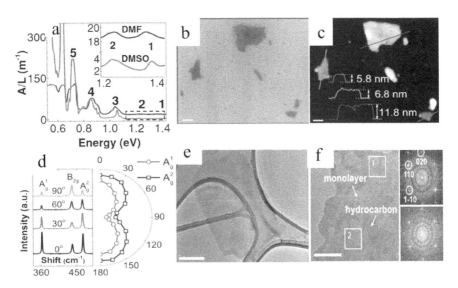

FIGURE 3.8 Optical and morphology analysis of BP crystal using sonication-assisted organic solvent exfoliation. a) The optical absorption spectroscopy employed to examine the nanoflakes in the range of 830 nm to 2400 nm (near IR) in particularly two solvents (DMF and DMSO). Interestingly, it was found that many spectral peaks appeared in the NIR range at around ≈1.38 eV, ≈1.23 eV, ≈1.05 eV, ≈0.85 eV, and ≈0.72 eV for both of the solvents which were due to the absorption of thick BP nanoflakes of mono to five layers. b&c) scanning electron microscopy (SEM) and AFM carried out the detailed study of morphology and thickness distribution of the layers. DMSO flakes of thickness around 15–20 nm were found while flakes were thinner than 5 nm in DMF. d) Modes of A1g and A2g after rotating the sample at 180°. By keeping in view the structure of BP, Ag modes were at the maximum level when laser polarization was parallel to the X-axis of the crystal. From a single flake in various orientations, the four spectra obtained are shown in the left image. The orientation-dependent intensities peak A_1g and A_2g are shown in the right image. The strong variation in the peak intensity corresponds to the anisotropic crystalline structure of the exfoliated flakes. e) The crystallinity, quality, and several layers BP nanoflakes were analyzed due to TEM images (scale bar = 200 nm). f) TEM image and FFT (whole flake exhibit similar features that propose even existence of single-layer BP) FFT in selected areas suggested BP monolayers, nanoflake, single crystalline, over the whole image. Adapted with permission from Yasaei et al. (2015), Copyright © 2015, WILEY–VCH Verlag GmbH & Co. KGaA, Weinheim.

yellow (Figure 3.9d). Therefore, it was found that the efficient solvent exfoliation with DMF (37.1 mJ/ m²) and NMP (40 mJ/ m²) having the value of surface energies which are near as the exfoliation requirement criteria of 2D BP (35–40 mJ m⁻²). The main benefits of DMF and NMP based exfoliation techniques such as scalable, high yield BP based nanomaterial, etc., that might be responsible for enhancing potential utilization as a building block for high-performance BP-based devices require a large amount of product (Coleman et al. 2011; O'Neill et al. 2011; Ren et al. 2017; Sun et al. 2017).

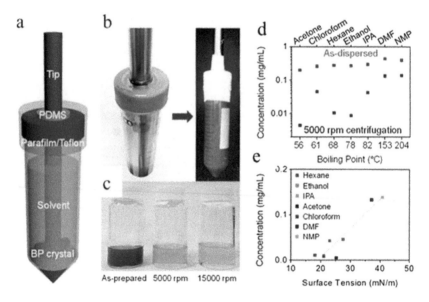

FIGURE 3.9 Optimal solvent for BP exfoliation via tip ultrasonication. a–c) The effect of the solvent was explored on BP exfoliation by keeping other experimental conditions constant. For safety purpose, moisture-free conditions, without O_2 and H_2O were adopted. a) Schematic representation. b) Snapshot of ultrasonication assembly setup (during processing having a decreased coverage to ambient air). c) From left to right after ultrasonication, without centrifugation (as prepared) and with different centrifugation speeds such as 5000 rpm and 15,000 rpm centrifugation speed, the snapshot of dispersion of BP is prepared in NMP as a solvent. d) The concentration plot of BP dispersion for various solvents such as acetone, chloroform, hexane ethanol, isopropyl alcohol, DMF, and BMP with various boiling points before and after treating the sample to 5000 rpm centrifugation speed, and e) after 5000 rpm centrifugation, there is variation in the value of surface tensions. Adapted with permission from Kang et al. (2015), Copyright © 2015, American Chemical Society.

It was found that by raising the boiling point and surface tension of solvents, the BP concentration monotonically increased (Figure 3.9d–e); a similar trend was observed in the case of graphene (O'Neill et al. 2011). During the analysis, a stable BP dispersion was found in NMP solvent while DMF was found as second highest. Microscopy and spectroscopy techniques were utilized to characterize the BP exfoliation in NMP (Brent et al. 2014). In particular, for SEM and AFM analysis, a very minute quantity (a droplet of fine BP dispersion) was put on the 300 nm SiO_2/Si substrates for ~5 minutes, blown off with N_2 gas to remove the liquid portion and for the uniform thin film on the surface of the substrate. After that sample was annealed on a hot plate for ~2 minutes at ~70°C. AFM was employed to measure the thickness values range (16 to 128 nm) (Figure 3.10a). SEM and AFM results were found agreed well regarding the morphological aspects of BP flakes (Kang et al. 2015). For other atomic-scale characterization such as for TEM and HRTEM analysis, BP flakes as a

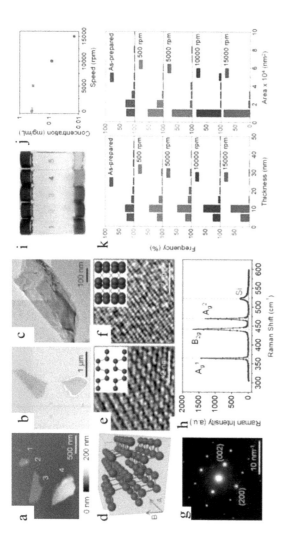

FIGURE 3.10 Characterizations of BP nanosheets using solvent-exfoliation technique. a) AFM show the surface morphology and height of BP nanosheet obtained from solvent-exfoliation. AFM was employed to measure the thickness values range (16 to 128 nm such as 1:16, 2:40, 3:29, and 4:128 nm). b) SEM image shows the surface morphology of BP nanosheet synthesized as a result of the solvent-exfoliation technique. c) The low-resolution TEM image shows the internal morphology of BP nanosheets synthesized as a result of solvent exfoliation. d) Schematic representation of the atomic structure of BP. HRTEM images show the BP nanosheet's internal morphology obtained as a result of solvent exfoliation along the direction. e) A. f) B. g) SAED pattern shows the crystallinity of BP nanosheets obtained due to solvent-exfoliation. h) BP nanosheets obtained as a result of solvent-exfoliation are characterized by Raman spectrum. i) Thickness, concentration and lateral area distribution with various centrifugation resolution per minute speeds of solvent-exfoliated BP. Following different centrifugation conditions/speed (1 as-prepared, 2 obtained as a result of 500 rpm, 3 obtained as a result of 5000 rpm, 4 obtained as a result of 10,000 rpm, and 5 obtained as a result of 15,000 rpm centrifugation speed) and all the dispersion for centrifugation were made by dispersing the BP in NMP solvent. j) From as a result of part i) the different quantity of the five different dispersions in NMP solvent. k) From the result of part i) the measurement of the area histograms and thickness of the five different dispersion of BP in NMP obtained from AFM images. Adapted with permission from Kang et al. (2015), Copyright © 2015, American Chemical Society.

result of solvent exfoliation were found put onto the thin carbon film based TEM grids (for sampling) and the dry sample was treated for analysis. The TEM picture of a representative BP nanosheet is represented in Figure 3.10c. Figure 3.10d shows schematically BP nanosheet crystal structure, enlightening that A and B have high-symmetry directions. Figure 3.10e–f shows the HRTEM images of providing outlooks of these orders such as A and B, respectively, mentioned in the inset of BP structures.

Moreover, the SAED pattern (Figure 3.10g) confirmed the orthorhombic crystalline structure of BP. Additionally, it was found that there is a link between the rotation of centrifugation and the overall concentration of BP (Figure 3.10a–b). Moreover, the lateral size and thickness of flake are gradually decreased due to increasing the centrifugation speed, as analyzed from the AFM and shown in Figure 3.10c (Kang et al. 2015). In another study, Guo et al. (2015) have produced phosphorene with multiple properties and characteristics such as the excellent mood of stability in water and controllable dimensions and thickness of layers by using a basic-NMP liquid exfoliation method. Before sonication, around 30 mL saturated solution of NaOH/NMP was prepared, and after that, bulk phosphorus around (15 mg) was added to the solution and mixed well. The resulting mixture was treated for a sonication bath (40 kHz frequency) and conducts the liquid exfoliation of the bulk BP for four hours. Subsequently, four hours of sonication was performed (the exfoliation process might be completed) for removing any remaining non-exfoliated part of bulk BP at 3000 rpm for ten minutes. Lastly, for preparation of the sample for characterization for nanoscopic measurement (SEM, TEM, AFM), around a volume of 0.05 mL (of the thin and thick solution of water and phosphorene) of solutions/suspension were put on silicon with a surface area of 280 nm SiO_2 (square surface layer around 1 cm × 1 cm) and the dry sample was used for characterization.

To obtained different layers of thickness, exfoliated BP samples were centrifuged at different rpm. It was analyzed that a comparatively thick layer of single-layer BP/phosphorene (5–12 layers, called 12,000 phosphorene) can be separated from the supernatant type liquid of NMP solvent, obtained as a result of centrifugation when it was treated for centrifugation at 12,000 rpm for another 20 minutes. Also, dynamic light scattering, statistical AFM (Figure 3.11a) and TEM (Figure 3.11c) showed that the single-layer BP (phosphorene) could be obtained as a result of the centrifugation process at 12,000 rpm (designated 12,000 phosphorene) having a thickness range of around 5.3 ± 2.0 nm (for 5–12 layers) and average diameter of around 670 nm.

On the other side, thinner phosphorene (called 18,000 phosphorene, 2–7 layers) were separated from the supernatant applying additional centrifuged at 18,000 rpm for 20 minutes with an average diameter measured by AFM (Figure 3.11b) and TEM (Figure 3.11d) of about 210 mm and a thickness of 2.8 ±1.5 nm. Thus, it was found from the results that the thickness of phosphorene (a type of product) can be controlled by adjustable centrifugal speed (Guo et al. 2015). HR-TEM image (Figure 3.11e–f) and SAED pattern analysis of selected-area (Figure 3.11g) were used to study the phosphorene's crystallinity. Lattice fringes values of 3.23 Å (Figure 3.11e) and 2.24 Å (Figure 3.11f) are relevant to the (012) and (014) plane of the BP atomic layer crystal structure, respectively (Hultgren, Gingrich, and Warren 1935). Figure 3.11e–f shows the uniform lattices propose that the phosphorene structure manufactured due

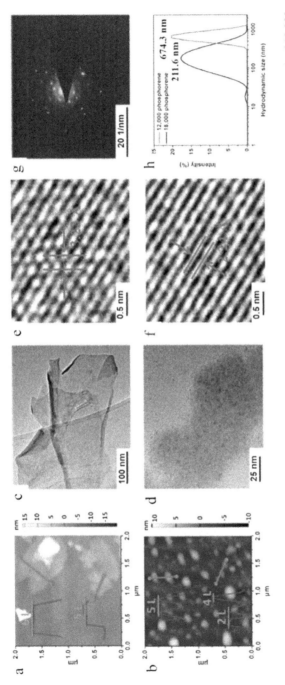

FIGURE 3.11 Height-mode of AFM images. a) Obtained as a result of 12,000 rpm centrifugation speed. b) Obtained as a result of 18,000 rpm single-layer BP. c) TEM images show the internal morphology of 12,000 single layers BP and d) 18,000 single layer BP. e&f) HR-TEM pictures show the d-spacing of single-layer BP with various crystal lattices. g) SAED pattern shows phosphorene's crystalline nature. h) Size distributions study of 12,000 and 18,000 SL-BP ascertained by dynamic light scattering (DLS). Adapted with permission from Guo et al. (2015), Copyright © 2015 WILEY–VCH Verlag GmbH & Co. KGaA, Weinheim.

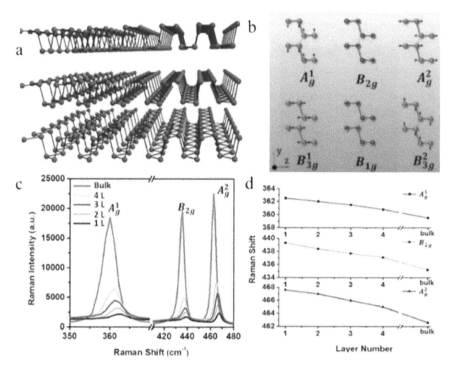

FIGURE 3.12 a) Illustrating the quadrangular pyramid structure of monolayer phosphorene using the concept of the group theory and conservation of momentum. Height-mode of 1–4. Left-hand side is single layer BP. b) Illustrating that Raman has six active modes with 12 lattice vibrational modes, but only three vibrational modes A1g, B2g, and Ag2 can be detected when the incident laser is vertical to the layered phosphorene plane. c) Shown the three Raman peaks having the values at 362.5 cm⁻¹, 439.3 cm⁻¹, and 467.6 cm⁻¹ from SL-BP having A1g, B2g, and Ag2 modes were detected, respectively. Adapted with permission from Guo et al. (2015), Copyright © 2015, WILEY–VCH Verlag GmbH & Co. KGaA, Weinheim.

to basic-NMP-exfoliation preserves the original crystalline state structure of the BP monolayer. Moreover, Figure 3.12a illustrates the quadrangular pyramid structure of monolayer phosphorene using the group theory and the conservation of momentum. Figure 3.12b illustrates that Raman has six active modes with 12 lattice vibrational modes. Still, only three types of vibrational modes—Ag¹, B₂g, and Ag²—can be detected when the incident laser is vertical to the layered phosphorene plane.

Three Raman peaks having the values at 362.5 cm⁻¹, 439.3 cm⁻¹, and 467.6 cm⁻¹ obtained from phosphorene having A1g, B2g, and Ag2 modes were detected, respectively, and are mentioned in Figure 3.12c. The Raman results of A1g, B2g, and Ag2 showed red-shifted upon increasing layer number; thereby, it provides a real *in situ* technique to govern the thickness value of phosphorene (Guo et al. 2015) (Figure 3.12d).

3.2.1.2.2   *Solvothermal Synthesis Method*

For the production of well-defined, monodispersed, and high-quality nanocrystals of nitrides, metal oxides, and novel semiconductor-based materials, a solvothermal-based reduction route is extensively employed due to its advantages, such as high pressure occurring in a pressure vessel. In particular, by utilizing this solvothermal heat treatment, nano-crystallites can be achieved with narrower size distributions and a higher degree of crystallization than conventional oil-bath heating. For solvothermal synthesis, normal solvents such as alcohols or water can be heated in a pressure vessel to temperatures far above their normal boiling points (Yang et al. 2007).

Xu et al. (2016) developed a robust and facile one-pot solvothermal method to prepare ultra-small BP nanoparticles (BPNPs) with an approximate size in the range of 2.1 ± 0.9 nm after using N-methyl-2-pyrrolidone (NMP) as a solvent and NaOH as a stabilizer. Different steps were involved during the synthesis of BPQDs, including: 1) the required crystals of BP first converted the BP into amorphous shape powders, then 2) the powdered BP was transferred into a reaction flask with the required combination of NaOH/ NMP solution, and finally 3) the solution was heated at 140°C with continuous vigorous stirring and under the influence of the nitrogen-based atmosphere for six hours.

The choice of solvent is very important for the synthesis of BPQDs. Among various solvents, NMP with a combination of NaOH greatly impacts the exfoliation process of 2D BP materials. Moreover, it was reported that BP easily gets the oxidized state result of irradiation under the visible light and is sensitive to water and oxygen. This is mainly because the combination of the saturated NaOH/NMP solution was selected for BPQDs and OH ion has great role for stability (Zhao et al. 2016; Ziletti et al. 2015), and all the experiments for the preparation BPQDs were carried out under the influence of a nitrogen gas-based atmosphere. After the required time for reaction, to separate the centrifugate and supernatant, the suspensions obtained from centrifugation were treated for centrifugation at 7000 rpm for 20 minutes. After centrifugation, the resultant supernatant (with the light-yellow color) represents the solution of NMP having BPQDs. Figure 3.13 shows the basic steps involved in the synthesis process

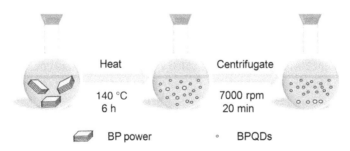

FIGURE 3.13   The basic experimental set up shows the synthesis process of BPQDs by the solvothermal method in the NMP/NaOH solvent system's presence. Adapted with permission from Xu et al. (2016), Copyright © 2016, WILEY–VCH Verlag GmbH & Co. KGaA, Weinheim.

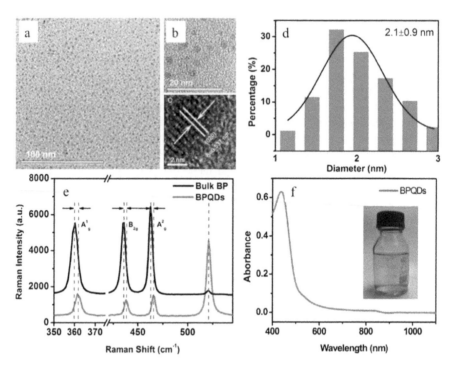

FIGURE 3.14 Characterizations of BP quantum dots (BPQDs) prepared by solvothermal process. a) TEM image. Shows the internal morphology of BP quantum dots. b) High-resolution TEM image with individual BPQDs. c) HRTEM pictures represent the d-spacing around 0.537 nm. d) Size distribution of around 100 BPQDs (using the statistical analysis) determined by TEM. e) Raman spectra show the different mood of vibrations exactly shows the presence of BP. f) BPQDs and NMP solution shows the linear absorption spectrum. Adapted with permission from Xu et al. (2016), Copyright © 2016 WILEY–VCH Verlag GmbH & Co. KGaA, Weinheim.

using the solvothermal method. The size and shape of the synthesized BPQDs are confirmed after properly analyzing the TEM images.

The TEM picture in Figure 3.14a–b shows the average size (2.1 ± 0.9 nm) of BPQDs, as shown in Figure 3.14d. In the HRTEM image of BPQDs, the above-mentioned lattice fringes with a fixed value around 0.57 nm were conforming to the (020) plane crystal structure of the BP shown in Figure 3.14c (Zhang, Xie, et al. 2015). The Raman spectra show the three different peaks, consistent with A1g (out-of-plane, OOP) shaking mode, B2g (the vibration modes in in-plane direction) and Ag2 mode, as illustrated in Figure 3.14e. This evidence strongly recommends that the average value of the thickness of BPQDs during the solvothermal process is pretty thin compared to other methods (Ziletti et al. 2015). The linear absorption spectrum of BPQDs is shown in Figure 3.14f. Consequently, this solvothermal synthesis is easily approachable to make it possible to synthesize BPQDs on a large scale compared with other previously reported methods such as liquid exfoliation and mechanical methods.

3.2.1.2.3 Surfactant-Assisted Exfoliation

There are exciting characteristic features of nanomaterial if there is a controlled shape of morphologies at the nanometer scale. Surfactants are behaving as a unique class of surface-active (such as carbohydrates, proteins, surface-active polymers, and small organic molecules) and possess a tremendous ability to control the crystal growth in a morphologically (size and shape) controlled manner. As a result of controlling the self-assembly behavior and its surfactant architecture, the fine-tuning of the desired morphologies can be achieved. Surfactants play a key role in the adsorption of surface-active moieties on different nucleating centers' crystal planes. They act as shape-directing agents in the synthesis of nanomaterials.

Brent et al. (2016) reported the exfoliation process to a few-layer BP (FL-BP) using some surfactant or capping agent (1% w/v Triton X-100 (TX-100, $C_{14}H_{22}O(C_2H_4O)$n where n = 9–10) as a surfactant in water. The surfactant solution having the approximate volume (~15 mL) and bulk BP crystal (~100 mg) were added in a reaction flask (flushed with argon), and the resultant suspensions were treated for sonication at a temperature of 298 K in an ultrasonic bath (820 W, for four hours) operating frequency is around 37 kHz and 30 percent power. The resultant suspension after 36 hours was treated for centrifugation at a rotating speed of 1500 rpm for a time duration of 45 min and the resultant liquid in the form of supernatant is removed and FL-BP of size around sub-20 nm thickness and 100~200 nm inside were obtained from the developed method (Kumar et al. 2016).

In another investigation, Kang et al. (2016) introduced a few-layer phosphorene (FL-BP) production in aqueous media using the surfactant-assisted ultrasonic exfoliation approach. According to the approach, the water is first degassed, and after that mixed, the surfactant (2 percent w/v sodium dodecyl sulfate (SDS) and layered bulk black phosphorus (hydrophilic) were mixed in trough ultrasonication. As a result, stable FL-BP were produced, similar to other hydrophobic 2D nanomaterials (Lotya et al. 2009; Kang et al. 2014; Nicolosi et al. 2013; Green and Hersam 2009; Zhu et al. 2015). To explain this apparent paradox, freshly cleaved flat BP crystal was selected to measure the BP surface's hydrophilicity characteristics using the contact angle measurements (Figure 3.15a). It is clear that the contact angle of BP surface is approximately 57°, and that is a good representation for indicating the hydrophobicity of BP between the common layered material such as graphene oxide (GO) (~27°) and another 2D nanomaterial (~90°) including transition metal dichalcogenides and graphene (Kang et al. 2014; Wei, Lv, and Xu 2014). Moreover, the effect of surfactant SDS during the dispersion of BP was confirmed after conducting some controlled experiments; it was confirmed after the experimental approach that when the aqueous BP dispersion is prepared in SDS's presence, there is a stable dark-brown solution (Figure 3.15b, left). On the other hand, without the presence of surfactants, when the BP is dispersed in water, it precipitates quickly (Figure 3.15b, right). Additionally, further investigation was made from the resultant dispersions in Figure 3.15a (left), after the centrifugation process the supernatants liquid were carefully decanted and

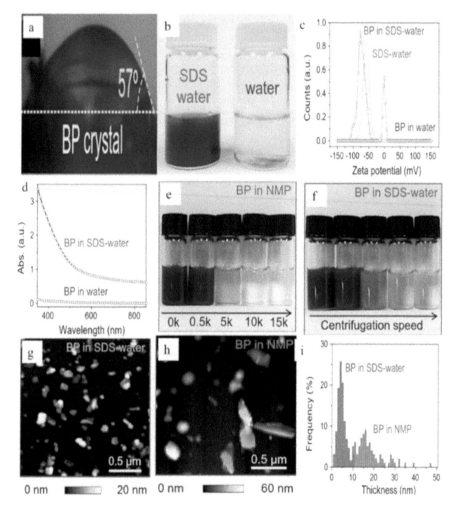

FIGURE 3.15 In an aqueous solution, the product of BP obtained as a result of the exfoliation process and its surface properties. a) Checking of hydrophobicity (measurement through contact angle) on an as-exfoliated flat surface of BP nanosheet. b) Snapshot of a BP based dispersion in water in the absence and presence of SDS. Zeta potential measurement of BP crystal in water, SDS in water, and BP crystal in SDS water. d) Optical absorbance measurement shows that BP's dispersions with SDS and without SDS. e&f) Micrograph of BP-based dispersions (without SDS) in NMP used as an organic solvent and BP-based dispersions in SDS water after the process of sonication and different speed of centrifugation machine (gradually increases from left to right in figure, as the arrow marked on bottles) at 500, 5000, 10,000, and 15,000 rpm. g&h) AFM show the surface morphology and thickness of exfoliated BP nanosheets treated in NMP and SDS water. i) Thickness distribution in BP nanosheets SDS water and NMP solvent. Adapted with permission from Kang et al. (2016).

further treated for zeta potential and optical absorbance measurements. The lower value of zeta potential value (Figure 3.15c) and higher optical absorbance value (Figure 3.15d), recommended the hydrophobic nature of BP nanosheets obtained as a result of exfoliation process. A stabilization process in aqueous solution with SDS similar to other 2D nanomaterial gets stabilized in aqueous solution in the amphiphilic presence (having both hydrophobic and hydrophilic groups) surfactant (Lotya et al. 2009; Nicolosi et al. 2013). Furthermore, this stability trend (of BP, as a result of liquid-phase exfoliation (LPE) in both SDS water) was also compared with organic solvent N-methyl-2-pyrrolidone (NMP). Additionally, both the reactions took place under the same exfoliation and centrifugation conditions. It was confirmed that the lighter yellow color obtained from BP dispersion in NMP indicates a lower concentration (Figure 3.15e) compared with the dark-brown solution prepared as a result of BP dispersion in SDS and water (Figure 3.15f) (Kang et al. 2015). Additionally, this evidence was further supported by optical absorbance values of the values of destruction coefficient used for the BP/NMP dispersion and the actual quantity of BP in a solution of SDS/water (Li et al. 2014). AFM images of both the dispersion, such as BP exfoliated in the presence of SDS-water (Figure 3.15g) and NMP (Figure 3.15h), confirmed the thickness through histogram measurement in Figure 3.15i.

Consequently, it is determined that the FL-BP nanosheets prepared in a solution of SDS-water have an average thickness (thinner) value (~4.5 nm) as compared with when BP dispersion is prepared in NMP (17.6 nm). These findings suggest the key role and contribution of aqueous solutions of surfactant for constructing thin FL-BP nanosheets when there is a comparison with organic solvents such as NMP.

3.2.1.2.4 Polymer-Assisted Exfoliation

Hybrid materials were obtained through different synthetic methods by dispersing BP nanoflakes in polymer matrices such as: 1) there is a process of mixing of FL-BP prepared as a result of liquid-phase exfoliation carried out as a result of preparation of poly-methyl methacrylate (PMMA) solution in organic solvent such as dimethyl sulfoxide (DMSO) and this process is termed as method A; 2) the direct approach used for the exfoliation process of BP carried out in a polymeric solution is termed as method B; 3) after completion the exfoliating process of BP in the liquid monomer such as methyl methacrylate (MMA), there is a process of *in situ* radical polymerization and it is termed as method C. The schematic detail of mentioned methods can be seen in Figure 3.16. There are two main benefits of these synthetic strategies, these approaches are helpful to enhance the process of BP exfoliation and at the same time the polymeric surfactant provides protection to the newly generated nanostructures material from oxidation when the BP nanoflakes are exposed to air and light (Passaglia et al. 2018).

The presence of both the phase, such as the polymer matrix, and the BP, were analyzed by different characterization techniques. According to size exclusion chromatography (SEC) analysis, the polymeric phase, such as the PMMA phase of the hybrids material, is prepared as a result of method A and B has the same or similar number average molecular weight ($\overline{Mn}$) when there is a comparison is carried out concerning its standard/reference (PMMA and PMMA-B-blank, respectively), while

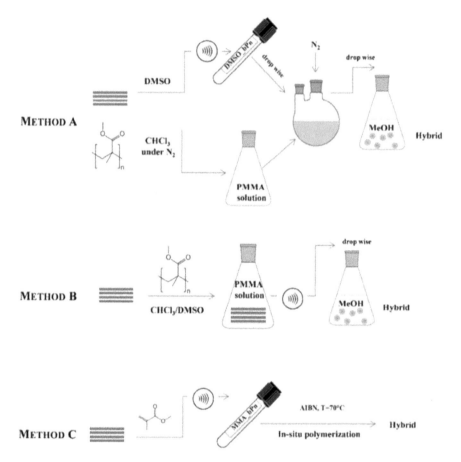

FIGURE 3.16 Schematic path of the procedures for the preparation of the hybrid materials. Adapted from ACS publication (Passaglia et al. 2018).

in case of method B there is a weak decrease of weight average $\bar{M}w$. On the other side, the sample preparation, as a result of *in situ* radical polymerizations (as a result of method C), the prepared sample was characterized by an extraordinary higher value (in the case of the average molecular weights, i.e., $\bar{M}n$ and $\bar{M}w$). These higher values of Mn and Mw in method C might be due to the presence of hindrance during the movement of the growing macro radicals, which might support the termination of reactions. As a result, there is an increase in the length of polymer chains (Nikolaidis, Achilias, and Karayannidis 2010). These findings strongly recommended for the growth of PMMA macromolecules between the layers of BP nanoflakes (BPn) near or possibly onto the surfaces of BPn, and the increase in growth of PMMA (increase the polymer chains) as a result promoted the growth of BPn. Furthermore, AFM analysis (as a result of the spin coating process used for the PMMA-BP-C anisole solution on glass sampling tablet) supported the evidence that strong van der Waal interactions between BPn and polymer chains. The presence of PMMA fractions

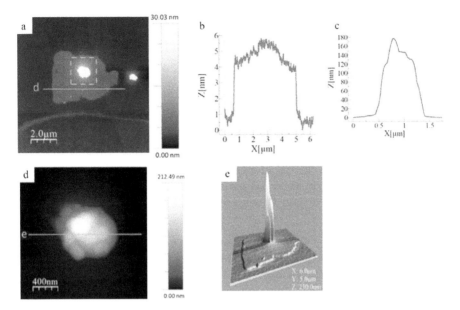

FIGURE 3.17 The reported hybrid of PMMA/BP. a) showed the visible "plateau" area is 4 nm higher value when compared with the surrounding PMMA thin film. b) High magnification image shows the square box in a) and represent the structure of BP around 4–5 μm wide. c) To represent the 3D region of interest from figure a), the BP nanoflakes up to 200 nm high and inhomogeneous. d) The zooming in and rescaling form of imagining showed the zooming in μm BP nanoflakes is an aggregated form of small size structures. e) The presence of height difference between the plateau and the BP aggregate shown in b), showed the presence of height difference between the plateau and the BP aggregate, approximately 200 nm. Adapted from ACS publication (Passaglia et al. 2018).

formed through thick aggregation is confirmed from Raman peaks of BP and the PMMA/BP hybrid is reported in Figure 3.17. This evidence strongly recommended an interaction between the portion of PMMA and BP flakes in sample PMMA-BP-C. It might be characterized by higher molecular weight polymerization of PMMA by generating P–C bonds (Passaglia et al. 2016; Li et al. 2016; Ahmed et al. 2017). Moreover, the feasibility of the *in situ* polymerization method and capable of efficiently exfoliating BP (protecting the BP nanoflakes) are further supported and verified using the different vinyl monomers such as N-vinylpyrrolidone and styrene. Consequently, BP and PMMA hybrids (BP nanoflakes are surrounded with PMMA matrices) are obtained, having a different variety of intriguing solubility, thermal, and mechanical properties (Abellán et al. 2017; Walia et al. 2016, 2017; Zhou et al. 2016).

3.2.1.2.5 Ionic Liquid Exfoliation

The facile, environmentally friendly, and large-scale liquid-exfoliation method is required to synthesize high-concentration and stable BP dispersions (from SL-BP) to FL-BP nanosheets nanoflakes arranged from the bulk BP as a result of the ionic

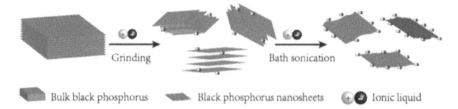

FIGURE 3.18 Schematic representation of green IL-exfoliation used to convert bulk BP into few-layer BP nanosheets. Bulk BP around 30 mg quantity was ground into small uniform-sized pieces in the presence of ILs (0.5 mL) for a time duration of 20 minutes; the process might be possible using a mechanical mortar and pestle. Adapted with permission from Zhao et al. (2015), Copyright © 2015, American Chemical Society.

liquids method. Zhao et al. (2015) developed an environmentally friendly, scalable, and facile method known as liquid-exfoliation using ionic liquid (ILs), a common IL ([BMIM][TfO]) and a functionalized IL ([HOEMIM][TfO]). Figure 3.18 showed that bulk BP around 30 mg quantity was ground into small uniform-sized pieces in the presence of ILs (0.5 mL) for a time duration of 20 min, and the process might be possible as a mechanical mortar having a pestle. Additionally, the mechanical shear forces produced as a result of the experimental set up are responsible for significantly minimizing the exfoliation time. It is possible that, as a result, the surface area of chunk BP extended toward magnitude (Yasaei et al. 2015; Zhang, Xie, et al. 2015; Zhang, Wang, et al. 2015). Consequently, after finishing the sonication process (used a mild sonication machine of around 100 W power, using an ice bath for a time duration of 24 hours), the resultant suspensions from monolayer phosphorene to FL-BP nanoflakes and nanosheets obtained from a bulk crystal of BP are formed as the final product material. Centrifugation process is carried out (at rotating speed around 4000 rpm for 45 min) to remove the un-exfoliated part of the BP structure and the inert form of supernatants solution having the thin layer of BP nanosheets (monolayer phosphorene to few-layer (FL-BP)).

To prove that the resultant mixture gained in both [BMIM][TfO] and [HOEMIM] [TfO] ILs is possible as a result of the exfoliation process of the bulk BP, and finally, the resultant product is in the form of monolayer phosphorene to few-layer (FL-BP) in ILs. To confirm the statement, a series of characterization as representatives can be conducted (left in Figure 3.19a). The successful BP exfoliation into monolayer phosphorene to few-layer BP (FL-BP) was confirmed by Raman spectra, as shown in Figure 3.19b.

Peaks of BP nanoflakes as a result of exfoliation in ILs such as [HOEMIM][TfO] IL at (~363, ~437, ~466 cm^{-1}) and [BMIM][TfO] at (~360, ~437, ~466 cm^{-1}), the mentioned vibrations peaks are related to the A1g, B2g, and A2g modes of FL-BP. The resulting modes showing the crystallinity of the BP atomic layers prepared a result of mechanical exfoliation (Saito and Iwasa 2015; Late 2015). These mentioned vibrational modes such as A1g, B2g, and A2g obtained as a result of IL-exfoliation, the resultant few-layer BP nanosheets show slightly blue-shifted when compared to the bulk BP.

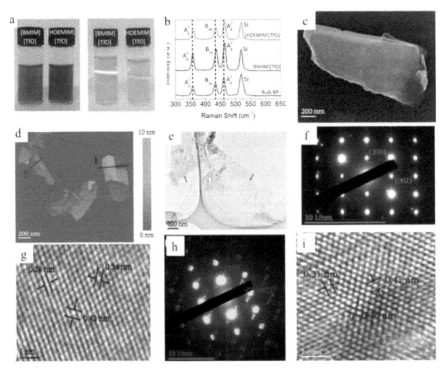

FIGURE 3.19 Optical characterization and morphology measurement of IL-exfoliated few-layer BP nanosheets. a) Snapshot of thin few-layers, the resultant mixture gained in both [BMIM][TfO] and [HOEMIM][TfO] ILs (left) is possible as a result of exfoliation development of the bulk crystal BP and the Tyndall effect of thinned distributions (right). b) The successful BP exfoliation into monolayer phosphorene to FL-BP was supported through Raman spectra. The spectrum of BP nanosheet as a result of exfoliation in ILs such as [HOEMIM][TfO] IL at ($\sim$363, $\sim$437, $\sim$466 cm^{-1}) and [BMIM][TfO] at ($\sim$360, $\sim$437, $\sim$466 cm^{-1}), the stated vibrations peaks are related to the A1g, B2g, and A2g modes of FL-BP. c) SEM of BP nanosheets in [HOEMIM][TfO] ILs represents the size and shape of the synthesized few-layer BP nanosheets. d) The AFM results investigate the topographic size and shape of the synthesized BP layers, ultrathin in nature they are achieved in [HOEMIM][TfO] ILs. e) The TEM images results that might be helpful to supplementary illustrate the exfoliated process of BP nanosheets when [HOEMIM]-[TfO] ILs is used. f) SAED pattern approving the excellency of the single-crystal structure of the few-layer BP nanoflakes having the orthorhombic crystalline characteristics. g) The HRTEM image supports the findings, suggesting that the exfoliation of few-layer BP flakes in [HOEMIM][TfO] ILs have the atomic-scale uniformity and the obtained lattice fringes have the spacing of 0.28, 0.34, and 0.42 nm, which is equivalent with the structure of phosphorene. h) The related electron microscopy characterization of few-layer BP nanoflakes achieved in [BMIM][TfO] ILs and i) magnified HRTEM image of same crystal taken from a selective area with lattice fringe values of 0.28, 0.34, and 0.42 nm, identifying the existence of monolayers. Adapted with permission from Zhao et al. (2015), Copyright © 2015, American Chemical Society.

This evidence confirms that as a result of exfoliation of bulk BP, the resulting product is in the form of thin BP few-layer nanosheet present in ILs (Guo et al. 2015). Other nanoscopic characterizations such as SEM, AFM, and TEM were used to monitor the morphology and crystalline structure of a few later BP in ILs. SEM of BP nanosheets in [HOEMIM][TfO] ILs is mentioned in Figure 3.19c, representing the synthesized FL-BP nanosheets' size and shape. Additionally, it is very important to know about the thickness distribution data obtained from AFM images for measuring the exfoliation performance. The AFM results investigate the topographic size and shape of the synthesized ultrathin BP layers obtained in [HOEMIM][TfO] ILs, shown in Figure 3.19d.

Consequently, different heights/thicknesses such as 3.58, 5.50, and 8.90 nm are observed from the AFM analysis, support the exfoliation process into FL-BP nanoflakes from bulk BP was successful (Wang, Yang, et al. 2015; Zhang, Xie, et al. 2015). Figure 3.19e showed the TEM images results that might be helpful to illustrate further the exfoliated nanosheets of BP when [HOEMIM][TfO] IL is used. The SAED pattern supports the refined and high-quality single-crystal structure of FL-BP nanoflakes having the orthorhombic crystalline characteristics (Figure 3.19f) (Wang, Yang, et al. 2015; Kang et al. 2015).

Figure 3.19g shows that the HRTEM image supports the findings, suggesting that the exfoliation of few-layer BP flakes in [HOEMIM][TfO] ILs have the atomic-scale uniformity and the obtained lattice fringes have the spacing of 0.28, 0.34, and 0.42 nm, which is equivalent with the single-layer crystal structure of BP. Figure 3.19h (Yasaei et al. 2015) shows the related electron microscopy characterization of few-layer BP nanoflakes/nanosheets obtained in [BMIM][TfO] ILs, again confirming the findings as mentioned earlier as in case of [HOEMIM][TfO] ILs (Figure 3.19i) (Zhao et al. 2015).

3.2.1.3 Electrochemical Exfoliation

During the past decade, the process of electrochemical-based exfoliation (ECE) has been magnificently engaged for 2D layered material such as graphene (Parvez et al. 2014; Ambrosi and Pumera 2016), MoS_2 (You et al. 2014; Liu, Kim, et al. 2014), and Bi_2Se_3 (Bi_2Te_3) (Ambrosi, Sofer, and Pumera 2017; Ambrosi et al. 2016). The use of this approach, mostly used for graphene, obtains outstanding findings in expressions of the cast, yield, and superiority of the exfoliated nanosheets. Technology advancement is required to develop a simple, robust ECE method in aqueous media to produce few-layered BP sheets starting from bulk crystals (Ambrosi and Pumera 2016; Ambrosi, Sofer, and Pumera 2017). Electrical chemical exfoliation is further divided into two main categories: anodic exfoliation and cathodic exfoliation, which are discussed in the following sections.

3.2.1.3.1 Anodic Exfoliation

Erande et al. (2015) used high-quality bulk BP crystal dispersed in ionic liquid solution (Na_2SO_4 (0.5 M in water)). The resultant mixture was transferred to a conducting stainless steel container, which was employed as a conductor and a basic foundation for electrochemical exfoliation. On the other side, the platinum wire (PtW) was introduced as another electrode named the counter-electrode. Both the electrodes were kept in a parallel direction by maintaining a fixed distance of around 0.5 cm between the working electrode and the counter-electrode. Moreover, it is a basic requirement to start an electrochemical process because there should be a limit of positive bias voltage value around +7 V and to the working electrode a current of about 0.2 A is needed. Consequently, the number of gas bubbles generated on the side of the platinum (Pt) electrode (H_2 gas) and meanwhile transferred to the working electrode (O_2) (continuously flow for 50 minutes) were used for the estimation of the rates of reaction. The resultant product obtained in the form of exfoliated BP nanosheets like morphology was collected in a beaker, which is kept under inert conditions for two hours to develop the precipitate. To get a clear product in a purified form, the as-synthesized raw precipitate was washed several times using distilled water and after that, the product was cleaned using ethanol solvent and dried in an oven at 60°C for three hours.

Other nanoscopic characterizations such as emission scanning electron microscopy (FESEM), AFM, optical microscopy (OM), and TEM were used to monitor the morphology and crystalline structure of an FL-BP. The electrochemically exfoliated BP sample shows the morphology (sheet-like) having a size ranging from 5 and 10 μm as shown in typical optical images (Figures 3.20a). On the other hand, the AFM image was used to investigate the morphology of bi-layered nature and equivalent height profile (almost 1.4 nm) for the synthesized BP nanosheet, as shown in Figure 3.20b and 3.20c, respectively. The surface morphology (size ranging from 5–10 μm and having a thickness of around 1–5 nm) of the BP nanosheet was further analyzed by FESEM analysis, as shown in Figure 3.20d. The comparative Raman spectra of BP nanosheet obtained from the electrochemical exfoliation process and bulk BP crystals nanosheets are shown in Figure 3.20e and 3.20f. Interestingly, both the nanosheets and bulk crystals demonstration the characteristic peaks agreeing to the A1g, B2g, and A2g phonon type modes of oscillation creating from the out-of-plane (A1g) and in-plane (B2g and A2g) vibrational modes (Figure 3.20g–h). This evidence strongly recommended that after finishing the electrochemical exfoliation process, the BP nanosheets remain crystalline (Erande et al. 2015).

In another study, Ambrosi, Sofer, and Pumera (2017) synthesized BP crystals using red phosphorus as a basic stating material and applied the thermal conversion method to red phosphorus (RP) to get a BP crystal (Figure 3.21a). Figure 3.21a shows the bi-electrode system set up in a container including a Pt foil as the cathode and BP flake working as the anode (there is a connection with Cu tape). The electrode was kept in a 0.5 mol/L H_2SO_4 aqueous solution in a parallel locating with a fixed measured distance of 2 cm (Figure 3.21b). To facilitate the wetting, a preliminary positive value of DC voltage of ~1 V was provided to the targeted crystal for two minutes. After that, a fine and slow discharge of the material was detected from the crystal structure with a solution turning slowly to yellow/orange (Figure 3.21c) when the voltage was raised to +3 V.

Moreover, for collecting a sufficient amount of material, the exfoliation process was processed continuously for two hours (Figure 3.21d) at the fixed supply of voltage around +3 V. After that, the collected exfoliated material was washed

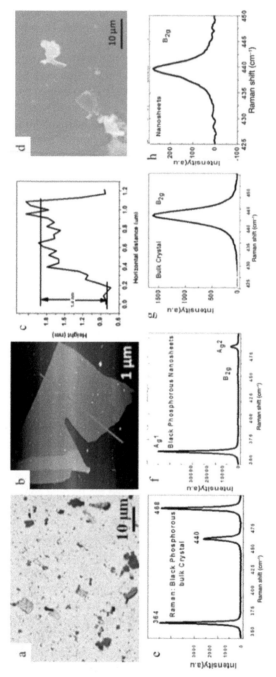

FIGURE 3.20 Surface morphology characterization of BP crystal: a) Optical microscope images show the electrochemically exfoliated BP sample showing the morphology (sheet-like) having the size ranging from 5 and 10 μm. b) Typical AFM picture indicated the surface morphology of BP crystal. c) The corresponding height profile of BP crystal. d) FESEM analysis showing the surface morphology (size ranging from 5–10 μm and having a thickness value of 1–5 nm) of BP nanosheet. e&f) The comparative Raman spectra of BP nanosheet obtained as a result of the electrochemical exfoliation process and bulk black BP crystals nanosheets. g&h) Both the nanosheets and bulk crystals display characteristic peaks equivalent to the A1g, B2g, and A2g phonon type modes of oscillation creating as a result of out-of-plane (A1g) and in-plane (B2g and A2g) vibrational type modes. Adapted with permission from Erande et al. (2015), Copyright © 2015, WILEY–VCH Verlag GmbH & Co. KGaA, Weinheim.

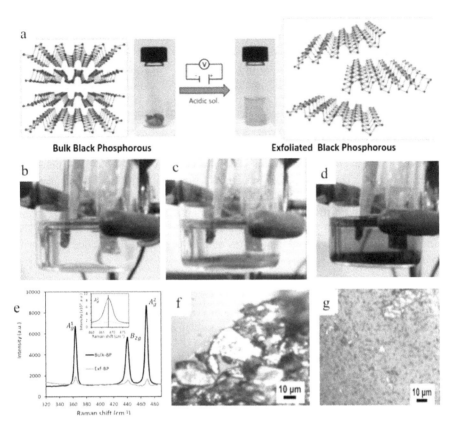

FIGURE 3.21 Anodic-based exfoliation method of BP. a) Schematic representation of procedure used for exfoliation. Layered type crystal structure of BP is obtained as a result of a DC voltage and is exfoliated in acidic aqueous solution. Pictures of the starting BP crystals (left) and the exfoliated material dispersion in DMF (right) are also shown. The photograph of the electrochemical setup shows the separation process of both the anode (BP flake) and cathode (Pt foil) in acidic solution (0.5 M H_2SO_4) by a fixed distance of 2 cm at b). In the absence of potential c) after the time duration of 20 minutes after using a voltage of +3V and d) after the passage of two hours' exfoliation process. e) The darker line represents the Raman spectra of BP crystal, and the lighter line shows the Raman spectra as a result of electrochemically exfoliated BP. Raman shift (between 460 and 480 cm^{-1}) is presented in a magnified form as an inset to show the blue-shift variation of the Ag^2 vibration mode of BP nanosheets as a result of the exfoliation process. Optical microscopic photos show the surface study of BP: f) surface study of bulk BP crystal and g) surface study of exfoliated BP nanosheets. Adapted with permission from Ambrosi, Sofer, and Pumera (2017), Copyright © 2017, WILEY–VCH Verlag GmbH & Co. KGaA, Weinheim.

plenty of times using ultrapure water through the process of vacuum filtration and after that, the resultant residue product was dried in a vacuum oven for 48 hours at 40°C. The method is reliable and can easily control the BP layer's thickness value, as demonstrated by Raman spectroscopy and STEM. The three vibration bands, A1g, B2g, and Ag2, are visible for both materials (bulk BP and exfoliated BP) as shown in the Raman spectra. It is clearly observed that the bands' intensity value in the case of the resulted exfoliated BP is decreased gradually. This evidence strongly indicates that the thickness of the BP layer is reduced (Castellanos-Gomez et al. 2014). Additionally, the number of the reduced layers is further confirmed by the blue-shift of the Ag2 band in case of the exfoliated BP layers (inset in Figure 3.21e) (Feng et al. 2015; Guo et al. 2015). The various structural topographies of the BP layers resulting from exfoliation and bulk crystal of BP are shown in optical images Figure 3.21f and 3.21g, respectively. Particularly, it can be noticed that as compared to bulk BP crystal as the starting material, the product of BP layers after the exfoliation method yields nanoflakes with minor lateral dimension and reduced thickness values (Ambrosi, Sofer, and Pumera 2017).

3.2.1.3.2 Cathodic Exfoliation

In this technique (Huang, Hou, et al. 2017), a Pt nanosheet (10 mm in width, 12 mm in length) was used as the counter-electrode and BP crystals (width 5 mm, 10 mm in length) were active as the cathode, and between the counter-electrode and working electrode, there is a distance of ≈1.5 cm. A DC power supply was introduced to maintain the static potentials in the range of −2.5 to −15 V to bulk BP electrodes. The electrolyte was prepared as a result of the desolation of tetrabutylammonium hexafluorophosphate (TBAP) with a quantity of 7.75 g, in dimethylformamide (DMF) with a volume of 40 mL. As a result of the sonication process for three minutes' duration, the exfoliated BP materials were completely discrete in an electrolyte solution. Consequently, anhydrous ethanol and DMF were used ten times to remove DMF and TBAP from prepared dispersion. At constant applied voltage, tetrabutylammonium cations were activated to get the addition into the interlayers of BP; as a result, there is a process of curling, stripping and peeling of BP (Figure 3.22). Also, different morphology monitoring techniques such as SEM, TEM, and AFM were used for monitoring the surface morphology of the phosphorene obtained at a voltage of minus 5 V (represented as FL-P-5), as shown in Figure 3.23a–d. Figure 3.23a shows the SEM images of sheets having large-area, which extended the micrometer-range dimensions. Additionally, TEM images showed the curly, ultrathin slices of phosphorene (Figure 3.23b). Moreover, an inter-planar space of 0.333 nm represents an equivalent value to the (021) plane of crystalline BP as further confirmed by HRTEM, as shown in Figure 3.23b (inset). AFM images (Figure 3.23c) and the equivalent thickness scattering chart are shown in Figure 3.23d. It is clearly observed that the value of thickness for phosphorene is distributed mainly between 0.76 and 0.79 nm resultant to approximately a combination of phosphorus atoms having two layers (Batmunkh, Bat-Erdene, and Shapter 2018).

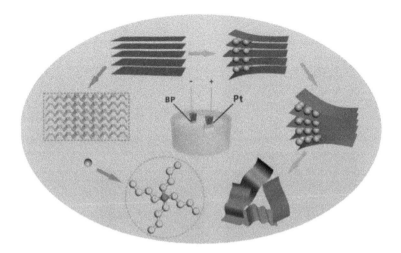

FIGURE 3.22 Electrochemical exfoliation process of phosphorene. Adapted with permission from Huang, Hou, et al. (2017), Copyright © 2017, WILEY–VCH Verlag GmbH & Co. KGaA, Weinheim.

3.2.1.4 Laser Irradiation

The laser irradiation method was used for the synthesis of highly crystalline BP nanosheets from the bulk crystal. Erande et al. (2015) initially used mortar and pestle for crushing around 13 mg of the bulk crystal of BP into small pieces in the presence of 5 mL of a solvent such as an isopropyl alcohol (IPA). After that, the sample was dispersed in IPA (having a quantity of ~ 0.5 mg/mL) and then transferred to a quartz beaker for laser treatment (with a fixed value of wavelength around 248 nm, spot size ~80 mm²). The process of irradiation was prepared with ports to sustain the inert (Ar-based) atmosphere. Moreover, to protect the BP crystallites and keep them from settling, the sample was continuously stirred during the process of laser irradiation. Consequently, after a one-hour irradiation process, the BP contents were allowed to settled at the lowest surface of the container and only the low mass/thin layer of BP sample contents were floating at the top. After that the whole sample was treated for centrifugation at 10,000 rpm rotation speed and then dried in a vacuum oven at 80°C for a time duration of four hours for further characterizations (Suryawanshi, More, and Late 2016).

The surface morphology of BP crystal before the process of laser irradiation is shown in Figure 3.24a, and Figure 3.24b–c shows the variation in surface morphology of BP crystal after laser irradiation.

Additionally, the thickness of the BP nanosheet layer (~7 nm) of BP nanosheet samples were carried out by AFM analysis, as shown in Figure 3.24d–e. Also, TEM exploration is used to attain a better internal morphological and structural representation of the as-prepared (laser-based exfoliated) BP sample as shown in Figure 3.24f–g.

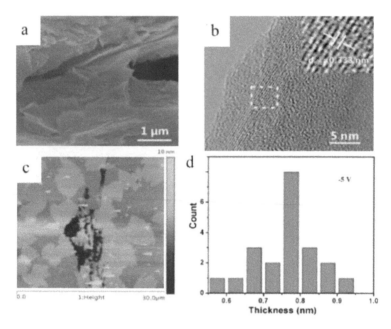

FIGURE 3.23 Different techniques used for monitoring the morphology of BP material. a) SEM is used for monitoring the surface morphology of BP nanosheet. b) TEM is used to monitor the internal morphology of BP nanosheet and, inset, HRTEM shows the d-spacing for BP. c) AFM pictures. d) The equivalent statistics of FL-P-5. Adapted with permission from Huang, Hou, et al. (2017), Copyright © 2017, WILEY–VCH Verlag GmbH & Co. KGaA, Weinheim.

Different morphology analysis techniques such as SEM, AFM, TEM, and Raman spectroscopy were introduced to investigate the morphological, structural, and optical representation of the synthesized BP sample. The SEM images show the surface morphology of the bulk crystal of BP, and the same model after the laser exfoliation process is presented in Figure 3.24a–c. For the sharp edges of the BP nanosheets, a higher magnification SEM image is recorded (Figure 3.24c) (Erande et al. 2015).

The SAED pattern describes the crystalline nature (six-fold symmetry) of the BP nanosheets (Figure 3.24h) and the HRTEM pictures of the BP nanoflakes show the d-spacing of ~0.25 nm, which relates to (221) plane as shown in Figure 3.24i. Figure 3.24j–k shows the Raman spectrum of pristine BP crystal displays well-defined peaks at 362.23, 437.07, and 464.75 cm^{-1}, corresponding to the phonon modes A1g, B2g, and A2g, respectively (Suryawanshi et al. 2016; Yeong, Maung, and Thong 2007; Late 2016; Wang, Jones, et al. 2015).

3.2.1.5 Thermal Annealing

To thin down the BP with extraordinary quality, selectivity, and poor energy utilization, a two-step method named thermal annealing is demonstrated by Fan et al. (2017). Additionally, the designed method in the first stage oxidizes the BP to P$_2$O$_5$ and then sublimates P$_2$O$_5$ instead of directly sublimating to BP, whereupon these

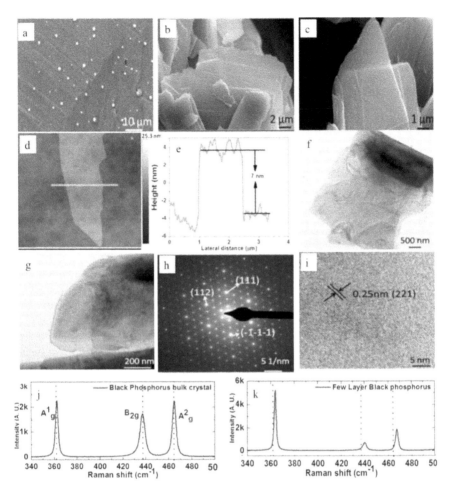

FIGURE 3.24 Surface morphology study of BP crystal as a result of the laser irradiation process. SEM photos of bulk crystal of BP: a) Dispersion of BP crystal before the process of laser irradiation and b&c) after laser irradiation the variation in surface morphology of BP crystal. d) Typical AFM image of few-layer BP nanosheets and e) corresponding AFM height profile of BP crystal. f) Low magnification bright field image to cover the broad area of the crystal. g) High magnification bright field image to cover the small area and watching the close dispersion of crystal structure. h) Selected area electron diffraction (SEAD) pattern and i) HRTEM image. j&k) Raman spectra of bulk crystal of BP and after the treatment of laser exfoliated BP nanosheets samples. Adapted with permission from the Royal Society of Chemistry (Suryawanshi, More, and Late 2016).

unique advantages are achieved. Both the stages require a shorter time and much lower temperature than sublimating the BP and other 2D layered materials. Moreover, the approach can achieve a stable and manageable thinning rate to produce a single layer BP with a high crystalline structure. It is possible after controlling the annealing temperature of P_2O_5 (330–360°C), which is about the range of sublimation temperature.

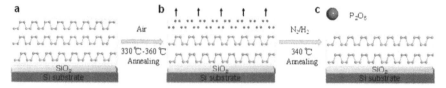

FIGURE 3.25 Thermal annealing method for thinning down the BP crystals. a) The exfoliation process of bulk BP flake after transferring onto a Si/SiO$_2$ substrate. b) The process of annihilation under 330–360°C in the open atmosphere helps the oxidation process of BP flake on the surface of top layers of the P$_2$O$_5$, and after further air annealing, P$_2$O$_5$ can be sublimated and, as a result, the obtained product is in the form of thinned flake of BP. c) There is an annealing process of BP flake under the influence of N$_2$/H$_2$ mixture at 340°C (maximum temp sublimation for P$_2$O$_5$) for maximum thickness. Adapted with permission from the Royal Society of Chemistry (Fan et al. 2017).

Figure 3.25a–c shows the mechanism involved for layer-by-layer thinning of BP and how the rapid thermal annealing is working. Figure 3.26a–b shows the resultant sample of bare BP and freshly exfoliated BP after treatment for AFM analysis. AFM measures the thickness of freshly exfoliated BP nanosheets, which are found to be around 0.6 nm thick (shown in Figure 3.26d), with the atomic layer of BP and the surface roughness of the monolayer BP measured as 2.6 Å, indicating its great surface quality as compared to bare BP flake (Figure 3.26c). The minimal surface roughness plays a key role in obtaining a value of high carrier mobility, and it might be achieved by reducing the rate of surface scattering. TEM measurements to a BP flake were performed further to determine the lattice integrity after the annealing process. Figure 3.26e indicates that after the thermal annealing, the crystallinity and atomic structure are well-maintained. Also, the selected area electron diffraction (SAED) patterns (Figure 3.26f) might help differentiate the diffraction spots on behalf of the (200), (101) and (002) planes of BP, and it is noted that the BP sample becomes thinner and moves toward monolayer structure as the intensity ratio increases gradually between the reflections of (101) and (200) spots (Fan et al. 2017).

3.2.1.6 Scalable Synthesis of 2D Black Phosphorus

2D layered material such as metal oxides, In$_2$Se$_3$, TMDs (MoS$_2$, WS$_2$, WSe$_2$), graphene, and hBN are essential to applications in various areas, including electronics, catalysis, photonics, and biomedicines. Although a large number of research reports have been published to improve the synthetic protocol for layered material, there still exists a gap between the academic research lab and industrial applications on a large scale due to the lack of available synthetic produce for BP nanocrystals in large quantities with the required quality for a particular application. In other words, still maintaining the good process of uniformity as well as precise controls and large-scale production of 2D layered material is a big hurdle for practical application in industry. To move from academic laboratory research to industrial applied applications, it is required to address the issue of scaling up a synthesis that has been commonly conducted in a batch format with controllable reaction parameters.

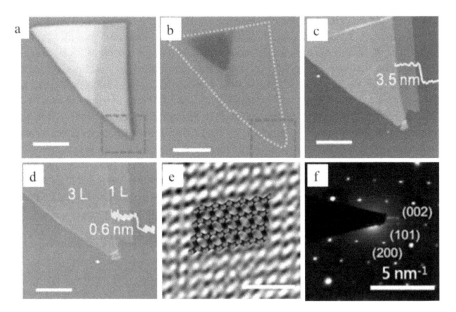

FIGURE 3.26 Optical microscopy images of BP crystals obtained as a result of thermal annealing. a) Simple BP nanoflake on 285 nm dimension Si/SiO$_2$ glass substrate. b) Representation of the reduced surface morphology of the same content of BP nanoflake after the process of rapid annealing process for 40 seconds in an open atmosphere and around 30 seconds in inert N$_2$/H$_2$ mixture flow under 340°C. c&d) The consistent AFM pictures with sale bar 5 µm of BP nanoflake-like morphology in the dotted line box of a) and c), respectively. e) HR-TEM image (scale bars: 1 µm.) of the same sample of BP after one minute N$_2$/H$_2$ annealing and two minutes air annealing and consecutively at 340°C (scale bars around: 2 nm). f) The analogous images of SAED pattern show the diffraction pattern of BP. Adapted with permission from the Royal Society of Chemistry (Fan et al. 2017).

Because the growth of colloidal nanocrystals is highly sensitive to experimental conditions such as reaction temperature, stirring rate, choice of solvent, reaction time, etc. Chen et al. (2016) have prepared high-quality BP crystals (size in nm range (5.2 to 9.4 nm)) obtained by a mechanical exfoliation that displays presently an on/off ratio of c. 5000 and mobility of c. 242 cm^2/Vs at 25–30°C for the preparation of few-layer samples. A quartz container (having a length of 13 cm and a diameter of 15 mm) was introduced, and AuSn alloy around (360 mg) quantity, SnI$_4$ (18 mg) and red phosphorus (900 mg), were first sealed and kept under lower pressure than the 10^{-3} bar. After that, the container was heated within one hour at 650°C by placing it horizontally in a tube furnace having a specific reaction zone. Moreover, after keeping it at 650°C for 24 hours, two different stages were applied; in the first stage, the container was treated for cooling to 500°C at a rate of 40°C/h, and after that kept at 500°C for at least 30 minutes for cooling. During the above-mentioned preparation methodology, large-size BP nanocrystals (with a quantity of 860 mg) were prepared on the container's sink side. It was noted that the reaction time is an important parameter, and on extending the reaction time, BP crystals were attained almost with the full conversion. Additionally, for getting specific and large BP crystals, the large crystal was removed and toluene and acetone were used as a solvent for washing to eliminate the residual contents/traces by the centrifugation technique. The hydrophilic nature of BP might be helpful for the mass production of BP nanosheets, and such synthesis is carried out by introducing the liquid-phase exfoliation process in water after using the pre-synthesized, highly refined BP nanocrystals as the preliminary materials during the reaction process. To improve the high-concentration dispersions (required for practical applications) of synthesized BP crystals, various parameters such as initial BP concentration (Ci), sonication time (ts), sonication power (Ps no more than 380 W), and centrifugation speed were checked on the BP nanosheets dispersed in water. It was analyzed that after keeping the fixed centrifugation speed (at around 2500 revolutions per minute (rpm)) there was an increase in the quantity of the distributed BP nanosheets linearly as a result of increasing Ps and ts as shown in Figure 3.27a–c. It was found that Ci had a great impact on the quantity of BP nanosheets when results were compared to other parameters mentioned in previous reports (Lange, Schmidt, and Nilges 2007; Nilges, Kersting, and Pfeifer 2008). Consequently, to produce an FL-BP nanosheet distribution of a reliable quantity, the BP crystals can be expeditiously exfoliated in water. Additionally, after retaining the high-quality bulk crystals, the BP nanosheets (BPNS) have a dispersion in water that is quite stable for further handling and technological applications. For practical applications, these BP nanosheets can be merged with graphene sheets (conductive) for the preparation of high-performance flexible paper-like lithium-ion battery (LIB) electrodes, which show an outstanding rate proficiency, sustained cycling efficiency, and the high value of specific capacity such as 501 mAh/g at a fixed value of current density around 500 mAg^{-1} (Chen et al. 2016).

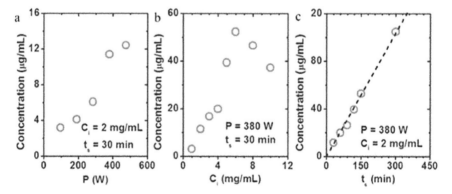

FIGURE 3.27 Few-layer BP nanosheets prepared as a result of the exfoliation process in water as a solvent. a–c) The effect of sonication power on a) shows the initial BP concentration, b) effect on sonication time. c) In water, the effect on the concentration of dispersed BP nanosheets. Adapted with permission from Chen et al. (2016) Copyright © 2015, WILEY–VCH Verlag GmbH & Co. KGaA, Weinheim.

Table 3.2 summarizes the different synthetic strategies of 2D BP nanomaterial, and the equivalent layers developed to improve the resultant product's quality (2D BP layered material) in the past decade. Among the developed synthetic protocols, liquid exfoliation represents a better and more reliable choice for large-scale production for BPNS. However, the structural damage caused by prolonging the sonication time and the small dimension of the isolated nanosheets are not suitable and might delay its use in electronic devices. Furthermore, other synthetic strategies, such as the tape exfoliation method for BPNS, are strictly limited to the treatment of 2D BP for electrical energy storage devices (EESDs). Similarly, due to the high cost, the pulsed laser deposition and laser irradiation are generally not very suitable for the scalable synthesis of 2D BP for EESDs. However, CVD and batch format approaches (synthesis of BP in the presence of polar organic solvent or water in a closed reaction vessel) are used as reliable, easily approachable, and efficient methods for producing high-quality 2DBPNS (Chen et al. 2016).

3.2.2 Epitaxial Growth

The bottom-up methodology depends on chemical reaction for the direct synthetic protocol of 2D BP layered material from various types of surfactant molecule or ligand (Wu, Pisula, and Müllen 2007; Cai et al. 2010; Li et al. 2009; Tian et al. 2018). In this section, we summarize the recent advancements in synthetic strategies applied for production, as well as the main advantages and shortcomings for future technological advancement.

3.2.2.1 Chemical Vapor Deposition

Advanced chemical synthesis techniques are used for 2D layered material such as TMDs and graphene. They might be provided with some inspiration for phosphorene synthesis, such as chemical vapor deposition (CVD) growth or hydrothermal synthesis (Castellanos-Gomez et al. 2014). Huang et al. (2016) stated that oxygen plays a key role in the degradation process of BP, and concluded that the single-layer BP might be stable in atmospheres/conditions such as oxygen-free water. For stability purposes, the boiling water might be provided to the environment, and in the case of hydrothermal synthesis, boiling water is used for the growth of stable BP. Smith, Hagaman, and Ji (2016) used the technique of vapor deposition to bulk red or BP for the development of a thin amorphous film of large size for red phosphorus.

For this process, they introduced CVD furnace under vacuum and grew the powdered red phosphorus on Si substrates as a result of heating at 600°C for 30 minutes to synthesize single and FL-BP thin films as large as 9000 lm^2 and having a thickness ranging from 3.4 nm to 600 nm (Figure 3.28a–b). Additionally, the further need of optimization for the other parameters (including pressure, temperature, reactant concentration, gas flow, growth time, etc.) plays a key role in a uniform size large-area few-layer BP (with average areas >3 μm^2). Thicknesses represent samples around four layers with average areas >100 μm^2. Moreover, Raman spectroscopy and transmission electron microscopy (TEM) have confirmed the successful growth of 2D BP from red phosphorus (Figure 3.28c–d) (Khandelwal et al. 2017; Smith, Hagaman, and Ji 2016). Large-area ultrathin 2D nanomaterials with excellent quality of the

TABLE 3.2
Summary of different synthetic strategies of 2D BP and the corresponding layers

Synthetic approach	Category	Sub Category	Starting material	Experimental conditions	Thickness	Size (nm/μm)	Ref
Non-epitaxial growth	Mechanical cleavage	Tap exfoliation	Block of commercial BP	Thin flakes were moderately peeled from a big block of commercial BP using scotch tape, 180°C residue collection temperature	Single-layer phosphorene	~0.85 nm	Chen et al. 2015; Liu, Neal, et al. 2014
		Ball milling	Red phosphorus	20 g of red phosphorus was introduced in a container under the influence of Ar atmosphere with a rotation speed of 400 rpm for 3.5 hours	Grain size	3–5 nm	Bao et al. 2018
		Wet chemical	2,2,6,6-tetramethylpiperidinyl-N-oxyl (TEMPO) and triphenylcarbenium tetrafluorobor ([Ph$_3$C] BF$_4$) are used as reactant or starting materials	Solvents such as dichloromethane (DCM) and a combined mixture of acetone and water with a fixed volume ratio (V: V=1:1) are used for thinning of BP in solution	Few layers	10.6 nm to 2.7 nm after chemical thinning	Fan et al. 2019
	Liquid phase exfoliation	Sonication-assisted organic solvent exfoliation	The chunk of the BP crystal concentration (~0.02 mg /mL) was dipped into various solvents, and the sonication technique for 15 hours with a total input power of approximately 1 MJ.	The exfoliation process generally consists of three different stages, including the dispersal of the material into the solvent, ultrasonication, and purification	Monolayer to a few layers	In the case when the DMF has used a solvent for exfoliation, >20% of the resultant BP nanosheet is thinner, less than 5 nm. On the other hand, when the DMSO has used a solvent, flake thicknesses in the range of 15–20 nm were obtained	Wu, Hui, and Hui 2018; Yasaei et al. 2015

Method	Starting material	Process	Product	Thickness/Height	References
Solvothermal synthesis method	Bulk BP crystals were powders into BP	In a reaction, flask added saturated NaOH solution in NMP solvent under vigorous stirring for 6 hours at 140°C under an inert gas atmosphere	Uniformly disperse BPQD	~0.573 nm	Xu et al. 2016
Surfactant-assisted exfoliation	Organic surfactant ~1% w/v Triton X-100 ($C_{14}H_{22}O(C_2H_4O)$ n where n = 9–10) as a surfactant in water. The inert surfactant solution in water (~15 mL volume) and bulk BP crystal (~100 mg) were added in a reaction flask (flushed with argon)	After ultrasonication, the resultant suspension after 36 hours was treated for centrifugation at a rotation speed of 1500 rpm for a time duration of 45 min and the resultant supernatant liquid detached	Few-layer black phosphorus (FL-BP)	Sub-20 nm thickness and 100–200 nm inside were obtained from the developed method	Brent et al. 2016; Kang et al. 2016
Polymer-assisted exfoliation	Bulk BP in a polymeric solution	The exfoliation process is carried out (in dimethyl sulfoxide, DMSO) with a polymeric surfactant such as poly-(methyl methacrylate) (PMMA) and indirectly mixed liquid monomer (methyl methacrylate, MMA).	Few layers	Height of the film ~200 nm	Passaglia et al. 2018
Ionic liquid exfoliation	3 mg/mL quantity of BP in Ionic liquid (ILs) such as [Emim] [Tf2N and [HOEMIM]-[TfO]) and the suspension is treated for moderate sonication (100 W) having an ice bath for the time duration of 24 hours	Exfoliated in [Emim][Tf2N] and exfoliated in [HOEMIM]-[TfO]	Monolayer to few-layer BP structure	3.58, 5.50, and 8.90 nm	Zhao et al. 2015; Wang, Yang, et al. 2015; Zhang, Xie, et al. 2015

(continued)

TABLE 3.2 Continued

Summary of different synthetic strategies of 2D BP and the corresponding layers

Synthetic approach	Category	Sub Category	Starting material	Experimental conditions	Thickness	Size (nm/μm)	Ref
Electrochemical exfoliation	anodic exfoliation	Bulk BP flakes	The stand electrode (Pt wire), a bulk crystal of BP, electrolyte (0.5 mol/L Na_2SO_4), with a required voltage (+7 V), and current value (≈1 mA)	3–15 layers	1–5nm	Erande et al. 2015
	Cathodic exfoliation	Bulk BP	Pt sheet and bulk BP is used as an electrode material, tetrabutylammonium hexafluorophosphate (TBAP) is used as an electrolyte in the DMF solvent, voltage ranging from –2.5 to –15 V	Thickness controlled layers synthesis (2–11 layers)	The thickness of the layers between 0.76 and 0.79 nm	Huang, Hou, et al. 2017; Batmunkh, Bat-Erdene, and Shapter 2018
Laser Irrigation	-	Around ~13 mg mass of the bulk crystal of BP divided into small pieces, convert into amorphous shape and afterwards used as starting material	Krypton fluoride (KrF) made laser having a wavelength of ~248 nm, with spot size ~80 mm², irradiation was equipped with ports to maintain an inert atmosphere	Few-layer	BP nanosheet (~7 nm)	Suryawanshi, More, and Late 2016; Suryawanshi et al. 2016; Yeong, Maung, and Thong 2007; Late 2016; Wang, Jones, et al. 2015
Thermal annealing	-	Bulk black phosphorus (BP)	Annealed the BP flakes under the flow of air and N_2/H_2 mixture	~7 layers	0.6 nm thick	Fan et al. 2017

Epitaxial growth	Chemical vapor deposition	-	Bulk red or BP	Under the controled conditions the phase transition of red phosphorus	~4 layers	3.4 nm	Smith, Hagaman, and Ji 2016
	Pulsed laser deposition	-	Bulk BP crystal	Graphene/copper substrates and bulk BP deposited at 150°C	Few layers	~2nm	Yang et al. 2015
	Molecular beam epitaxy		Bulk white Phosphorus	The process of BPQDs preparation were possible on fully deoxidized Si (111) and covered with native oxide Si (100) at 20°C and 15°C, respectively	Pyramid-shaped	Average radius of 27.5 ± 5 nm and height of 3.1 ± 0.6 nm	Xu et al. 2018

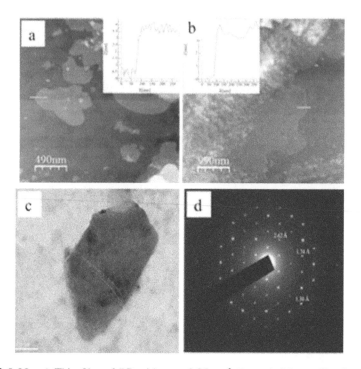

FIGURE 3.28   a) Thin film of BP with area 0.35 μm². Inset: height profile showing the thickness value of approximately four layers (3.4 nm) of BP thin film. b) The magnified AFM images of larger area content thin film of BP with respective inset part of the height profile that is also representative of four layers. c) The TEM pictures shows the small piece of BP sample after separating it from a silicon-based substrate and the respective d) SAED pattern images showing the crystalline nature of the BP. Adapted with permission from Smith, Hagaman, and Ji (2016), Copyright © IOP Publishing Image/photo/illustration.

crystal, tunable size, and thickness were produced by the CVD approach. Various types of 2D nanomaterials, such as metal oxides, In$_2$Se$_3$, TMDs (MoS$_2$, WS$_2$, WSe$_2$), graphene, and hBN, have been successfully prepared using CVD techniques (Qiu et al. 2017).

3.2.2.2   Pulsed Laser Deposition

Conventional pulsed laser deposition (PLD) is used to prepare ultrathin BP films at a temperature as low as 150°C, with a tunable direct bandgap. Despite extreme conditions such as high pressure and high temperature for the synthesis of thin-film from BP bulk (Zhang et al. 2012b; Hao et al. 2000). Yang et al. (2015) introduced the pulsed laser deposition (PLD) technique for phosphorus ultrathin films grown on SiO$_2$/Si or graphene/copper substrates where the distance between the substrates and the objective of bulk BP nanocrystal (smart-elements) is around 4 cm.

Additionally, the chamber's internal adjustment was furnished to a base pressure of around 1.5×10^{-7} Torr before PLD deposition. The temperature of the substrate was controlled at 150°C for synthesized a-BP thin films. The BP object was reduced using a Krypton fluoride (KrF) pulsed laser having the specifications such as ((wavelength λ = 248 nm) with 5 Hz replication rate). Further, there is a rotation process during the deposition process made by both BP target and substrates for achieving uniform film growth (Zhang et al. 2012a; Hao et al. 2000).

Consequently, the as-deposited ultrathin films were then stored in a high vacuum chamber and cooled down to room temperature naturally for characterization. The synthesized monolayer BP crystal film/SiO$_2$/Si was examined by TEM analysis to confirm the crystalline nature of the film using an approach called focused ion beam and a lift-off method. Figure 3.29a shows the low-magnification TEM (a very uniform and smooth film can be seen). Additionally, Figure 3.29b shows the HRTEM interface (a dense and disordered morphology) between the concerned substrate and grown film. Furthermore, SAED pattern images (inset of Figure 3.29b) indicate the amorphous nature of the obtained thin films.

AFM microscopy was used to investigate further the surface morphology and thickness (2 nm) of a-BP thin films (Figure 3.29c). Raman spectroscopy (Figure 3.29d) is used to characterize the allotropic behavior of the fabricated phosphorus

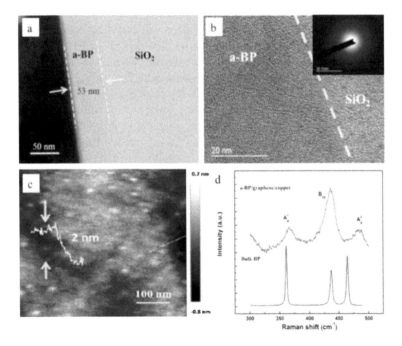

FIGURE 3.29 Surface morphology of BP crystals as a result of pulsed laser deposition. a) TEM images show the cross-sectional overview of thin films. b) HRTEM image showing the SEAD pattern in (inset), and representative powdered phases of the deposited films. c) AFM images show the thickness value of BP based thin films (c: 2 nm). d) Raman spectra of thin films deposited/placed at a temperature of 150°C and bulk crystal of BP grown on graphene/copper substrates. Adapted with permission from Yang et al. (2015), Copyright © 2015, WILEY–VCH Verlag GmbH & Co. KGaA, Weinheim.

films. Although the mentioned technique is used to synthesize BP crystals due to high cast, the pulsed laser deposition and laser irradiation are generally not very suitable for the scaleable manufacture of 2D BP-based nanomaterial.

3.2.2.3 Molecular Beam Epitaxy

During the past decade, molecular beam epitaxy (MBE) has been successfully employed to prepare different elemental 2D materials such as stanene, germanene, and silicene. 2D structure composition of BP possesses a wide range of fascinating physical properties—including doping, strain, bandgap tenability, in-plane anisotropy, high carrier mobility—depending on the thickness of synthesized BP layers (Batmunkh, Bat-Erdene, and Shapter 2016; Bullett 1985; Carvalho et al. 2016). Despite harsh experimental conditions (Li et al. 2015; Smith, Hagaman, and Ji 2016), the fabrication process of BP is rarely reported due to poorly controlled phase transition, mechanical and liquid exfoliations (Lee et al. 2016; et al. 2015b). The MBE method (Zhu and Tománek 2014) is successfully employed to synthesize blue phosphorene (allotrope of BP), but there is still a deficiency of such a kind of epitaxial growth method for BP. Xu et al. (2018) presented the MBE technique to prepare BP quantum dots (BPQDs) on Si substrates, using white phosphorus as a starting material. As a result of the MBE approach, the few-layer BPQDs were successfully fabricated directly on Si substrates due to cleavage and cracking of the precursor, i.e., white phosphorus starting material. The process of BPQDs preparation due to epitaxial growth was possible on fully deoxidized Si (111) and covered with native oxide Si (100) at 20°C and 15°C, respectively. The surface chemistry of the synthesized BPQDs was characterized by X-ray photoelectron spectroscopy (XPS), atomic force microscopy (AFM), and Raman spectroscopy with fingerprint region peaks of BP. Figure 3.30a shows the AFM image of uniform BPQDs with a pyramid-shape synthesized with an average radius of 27.5 ± 5 nm and height of 3.1 ± 0.6 nm on Si (111) substrates (Figure 3.30b–c).

Figure 3.30d shows the cross-sectional profiles along lines 1 and 2 in which the thickness of pyramid shape BPQDs and Si surface step height was illustrated. Additionally, it was found that the BPQDs on the non-deoxidized Si (100) with an average size of 8 and 14 nm and height between 1 and 2 nm were grown. It was concluded that both the factor surface steps and surface orientation greatly influence nucleation and growth of BPQDs on Si (100) face. X-ray photoelectron spectroscopy (XPS) was performed to confirm the successful synthesis of BPQDs to measure the elemental composition of BPQDs/Si using the MBE method. The XPS mentioned in Figure 3.30e shows the crystallization characteristic of BP. It is due to spin-orbit split doublets of 2p1/2 and 2p3/2 at 130.5 eV and 129.4, respectively exhibits by BPQDs/Si(111) (Wood et al. 2014; Woomer et al. 2015).

Additionally, there is a broad peak in XPS at 133.8 eV showing the presence of phosphate species (POx) as a result of oxidation of BPQDs under ambient conditions (Wood et al. 2014). Furthermore, Figure 3.30f shows the Raman spectrometer which revealed the fingerprint vibration modes of BPQDs against the strong background signal due to Si substrate. It was observed that the BPQDs have low optical absorption. In contrast, the monocrystalline shape silicon substrate has high Raman cross-sections. Another significant identification showed the presence of BPQDs onto the

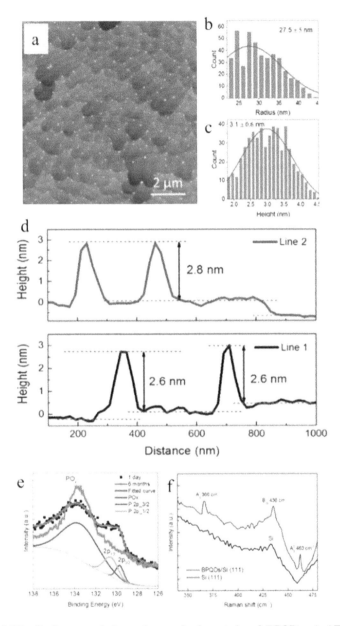

FIGURE 3.30 Surface morphology characterization study of BPQDs. a) AFM images of BPQDs grown on Si (111). b) Statistical size distribution study. c) Statistical height distribution study. d) Height profiles of mentioned lines 1 and 2, on AFM image a. e) showed the crystallization characteristic of BP and it is due to spin-orbit split doublets of 2p1/2 and 2p3/2 at 130.5 eV and 129.4, respectively exhibits by BPQDs/Si (111). f) Raman spectra peaks at 366 can be assigned to one out-of-plane phonon mode (Ag^1) while 436 cm^{-1} (B$_2$g), and 463 cm^{-1} (Ag^2) represent the two in-plane modes of BPQDs. Adapted with permission from Xu et al. (2018), Copyright © 2018, WILEY–VCH Verlag GmbH & Co. KGaA, Weinheim.

Si substrate, confirming the introduction of three peaks at 366 can be assigned to one out-of-plane phonon mode (Ag^1). At the same time, 436 cm^{-1} (B2g), and 463 cm^{-1} (Ag^2) represent the two in-plane modes (Eswaraiah et al. 2016; Lee et al. 2016). These findings provide a facile way for the synthesizing of BP from common allotropes of phosphorus, e.g., white phosphorus as a result of standard epitaxy systems. On applying the epitaxial growth approach, BP can be synthesized in scalable form and further extended the material toward functional heterostructures. Furthermore, the developed protocol improves the synthesis strategies in the form of the epitaxial growth other than exfoliation-based synthetic approaches of BPQDs and provides an insight and guidance for further fabrication of 2D BP, nanoribbons, nanosheets, or nanowires.

3.3   STABILITY PROPERTY AND DEFECT ENGINEERING OF BLACK PHOSPHORUS

In order to improve their use in practical applications, the goal of optimizing the chemical and physical characteristics of solids is to be optimized in the field of existing material science goals. By mixing chemistry, physics, and engineering in a clever way, researchers have been able to tune the initial characteristics of many materials. In conventional techniques for tuning material properties, certain external variables—such as friction, strain, and temperature—and intrinsic numbers—such as the density and shape of defects and dopants—are used. However, these methods have all been developed, updated, and adapted for bulk material so far. For 2D materials consisting of only one to three atomic layers, various strategies affecting the intrinsic quantities need to be improved and sometimes even unregulated, as they are no longer successful. To discover and account for ubiquitous defects and foreign atoms, 2D material science needs to resolve all aspects and develop successful strategies to monitor material properties by reliably introducing imperfections. The tuning of the intrinsic characteristics in 2D materials inevitably proceeds through atomic-scale modifications. Thus, 2D materials science deals with the phenomenon of defect engineering in its broadest meaning, i.e., authorizing the skillful addition, removal, or manipulation of atoms, and representing the key to open up the full potential for applications (Schleberger and Kotakoski 2018).

The simple BPQDs and a combination of BPQDs with PLGA (BPQDs/PLGA) composite NSs having a similar concentration of BPQDs (around 20 ppm.) were dispersed in water under an uncovered state to air for eight days to investigate the effect of PLGA encapsulation on the stability of BPQDs. Additionally, their optical properties were observed during the different time intervals (0, 2, 4, 6 and 8 days). The color of the solution containing the bare BPQDs become lighter during dispersion as shown in Figure 3.31a; on the other hand, there was no change in solution color of BPQDs/PLGA NSs and it stayed quite the same as before. Both the dispersions (simple BPQDs and BPQDs/PLGA NSs) are found in a stable state without sedimentation or visible aggregation, as the inset photographs show in Figure 3.31b–c, and it can be further reinforced by the Tyndall effect usually used for diluted type suspensions like material. Figure 3.31b–c show the corresponding absorption spectra. Both simple BPQDs material and combination of PLGA functionalized BPQDs composite NSs display a typical broad absorption band during initial dispersion time crossing the NIR regions and ultraviolet, while in the case of BPQDs the absorbance intensity is decreased in water with the passage of dispersion time (Shao et al. 2016). It is clearly observed from Figure 3.31b that the deviation in the ratios of absorption is found around 808 nm and associated with the innovative assessment (A0) and after the passage of two days the intensity (A) decreases by 27.5 percent and after eight days by 62.5 percent.

There is an irreversible reaction expected with oxygen (O_2) as well as with water (H_2O) and as a result an oxidized phosphorus species such as (P-PxOy) is formed. Meanwhile there is a transformation of PxOy to the final anions (that is, PO34) that is answerable for the degradation progression of BP in water (Huang et al. 2016). The degradation of the simple and bare BPQDs might be responsible for the decrease in absorbance. In contrast, the BPQDs showed lower absorbance than BPQDs/PLGA NSs, which might be due to the larger size of PLGA functionalized BPQDs as well as the contribution of the PLGA shells with BPQDs. Due to the durable hydrophobic characteristic of PLGA, the functionalized BPQDs with PLGA are effectively secure from water and oxygen and, as a result, under ambient conditions (standard temperature and pressure) the stability property of the PLGA functionalized BPQDs composite NSs in water is gradually enhanced. Moreover, there is no big difference observed in case of PLGA functionalized BPQDs NSs composites (Figure 3.31e) when compared with the Raman spectra (the blue-shift) of the BPQDs after dispersion in water for eight days (Figure 3.31d), confirming the better stability of BPQDs/PLGA NSs (Favron et al. 2015; Gan et al. 2015). Figure 3.31f–g further shows the temperature-based stability factor of the photothermal performance of both the bare/simple BPQDs and combination with a polymer such as BPQDs/PLGA composite NSs in water.

Initially after the treatment of 808 nm laser irradiation (1.0 W cm²) for a time duration of ten minutes the temperature value of the simple BPQDs solution is increased gradually by 19.3°C, but after the passage of time such as eight days due to the BP degradation the temperature enhancement is only 8.7°C. In contrast, in the case of composite PLGA functionalized BPQDs NSs the temperature enhancement of the solution is around 19.9°C after eight days and checked after irradiation treatment for a time duration of ten minutes, and it is quite near to the preliminary one of 21.1°C. This evidence strongly recommended that encapsulation of PLGA can effectively maintain the photothermal characteristics in water as a result of protecting the degradation of BPQDs.

In another study, Qiu et al. (2018) investigated that the BPNSs functionalized with polyethylene glycol–amine (PEG-NH_2) as a result of the electrostatic adsorption approach for improving the physiological stability. After completing the process of surface modification of the BPQDs there is variation in size from −28.2 mV to −16.7 mV respectively, the difference in the size of BPQDs was confirmed after using the zeta potential. The size distribution measurement (average diameters) study of bare BPNSs and after the functionalization with PEG is 155.6 nm and 160.3 nm. The size distribution is measured by dynamic light scattering (DLS) analysis, which showed a minor variation in dimensions afterward the process of functionalization of BPNSs (Qiu et al. 2018).

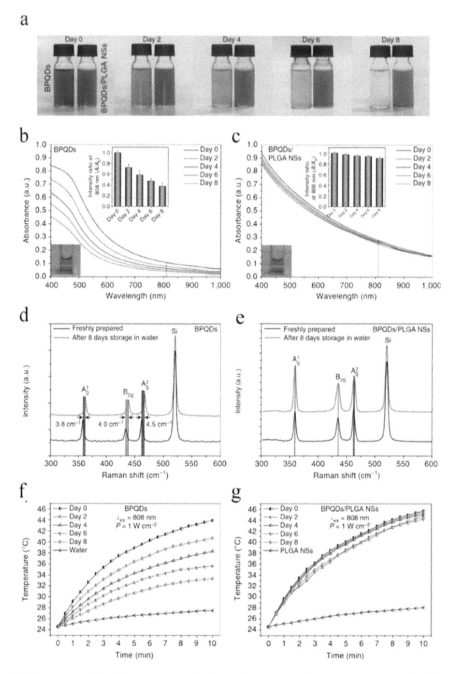

FIGURE 3.31 Stability property evaluation of black phosphorus quantum dot (BPQDs) under ambient reaction conditions. a) Snapshot of simple/bare BPQDs and BPQDs/PLGA during the passage of different time interval such as Day 0, Day 2, Day 4, Day 6, Day 8. b&c) Both the dispersions (simple BPQDs and BPQDs/PLGA NSs) are found in a stable state

FIGURE 3.31 Continued

without sedimentation or visible aggregation; the inset photographs show different absorption ratios (A/A_0) at fix wavelength 808 nm and Tyndall effect after different time interval such as Day 0, Day 2, Day 4, Day 6, Day 8. d&e) When compared with the Raman spectra (the blue-shift) of the BPQDs after there is a dispersion in water for eight days d), no big difference is observed in case of BPQDs/PLGA composites NSs e) substantiating the better stability of BPQDs/PLGA composite NSs. f&g) After the storage in water (simple BPQDs and combined BPQDs/PLGA) for different periods of time (Day 0, Day 2, Day 4, Day 6, Day8) and further showed the temperature-based stability factor of photothermal-based performance of both the bare/simple BPQDs and PLGA functionalized BPQDs composite NSs in the presence of water. Reproduced under the terms of the Creative Commons Attribution 4.0 International. Copyright © the authors, published by Springer Nature (Shao et al. 2016).

3.4 PROPERTIES AND APPLICATION OF 2D BP

BP has a layered structure similar to transition metal dichalcogenides (TMDs) and graphite, but the main difference is the unique puckered single-layer geometry. Due to variation in some of the parameters such as film thickness, anisotropic in-plane properties, and high carrier mobility, there is a variation in direct electronic narrow bandgap ranging from 0.3 eV to 2 eV of thin-film BP. Due to this wide-bandgap range, BP is an encouraging nominee for novel applied claims in the field of nanoelectronics and photonics with different from TMDs and graphene (Ling et al. 2015). Nowadays, BP is attracting a lot of attention due to its potential applications in optoelectronics and electronics (Li et al. 2014; Xia, Wang, and Jia 2014; Sun, Lee, et al. 2015; Yuan et al. 2015; Qiao et al. 2014; Dhanabalan et al. 2017; Song et al. 2016). That the anisotropic structure of black phosphorus is responsible for its anisotropic transport properties has been predicted both theoretically and experimentally, and these properties are responsible for transport management. Moreover, it was found that the thin layers exfoliating from bulk BP in a clean, atomically way can make it possible to extend its properties experimentally. Like other 2D layered material (2DLMs), a strain-controlled (Rodin, Carvalho, and Neto 2014; Cai et al. 2015) and thickness-dependent (Tran et al. 2014) bandgap originates in few-layer BP. Different properties were found in BP from monolayer to few-layers are discussed in the following sections.

3.4.1 BANDGAP

The thin film of BP (having the thickness value > 4 nm or the distribution of eight layers and having the moderate bandgap of BP (~0.3 eV)) can bond the value of energy gap between and the relatively large value of bandgap ranging of many TMDCs (1.5–2.5 eV) (76–79) and the zero bandgap of graphene (Novoselov et al. 2005; Zhang et al. 2005) for a diverse range of many important applications in the field of nanoelectronics and nanophotonics (Figure. 3.32a). Recently BP has been considered an attractive nominee for near and mid-infrared optoelectronics such as modulators, detectors, and potentially light-based generation devices like lasers and

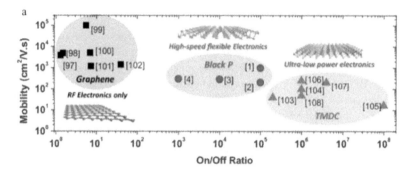

FIGURE 3.32 a) The mobility/on–off ratio value from the spectrum shows the performance regions of different 2D nanomaterial such as graphene, BP, and TMDCs (MoS$_2$, WSe$_2$, and WS$_2$) transistors. Adopted with permission from Ling et al. (2015).

light-emitting diodes (LEDs) due to the BP thin film having the strong optical conductivity in the range of 1–5 μm wavelength. Nowadays, imaging (62) and detectors (64) functions have been established using BP thin films. The wide tuning range of BP bandgap (as a result of the application of strain (5) and varying the layer number (80)) is considered the more attractive feature of BP for optoelectronics applications (Ling et al. 2015).

3.4.2 ON–OFF RATIO/MOBILITY

The transport properties of BP are between that of most TMDs and graphene. The frequency range and mobility value of the concerned material for transistors (on–off current ratio) is shown in transistor devices and mobility/on–off ratio spectrum of the different 2D materials, as shown in Figure. 3.33. Figure 3.33a represents the different frequency range and bandgap detail of different 2D layered materials mentioned in Figures 3.33b–e (Ling et al. 2015). Additionally, there are some key applications in each region of this spectrum, having the corresponding domains in the field of nanoelectronics. Due to the zero bandgap of graphene, the on–off ratio in the case of graphene containing transistors devices is usually not more than 10, but due to 2D semimetal characteristics, graphene shows very high mobility value. On the other hand, many monolayer TMDCs material, due to their relatively low carrier mobility value (in most cases less than 100 cm^2/Vs), have lately attracted much attention. Still, the value of on–off ratio value of TMDCs based transistors is quite high as above 108 with a high range of 1010 in some cases and having an application for ultra-low-power nanoelectronics.

It is not easy for TMDCs such as MoS$_2$ or graphene to cross the region of mobility/on–off ratio combination for BP on the plot. In this region, the device requires the on–off ratio in roughly the range of 103–105, and at the same time, the mobility value of the layered material ranging from a few hundred to 1000 cm^2/Vs. Due to the properties mentioned above, BP might be considered as an attractive and reliable candidate for the building of gigahertz frequency-based thin-film electronics (Qiao et al. 2014; Ling et al. 2015).

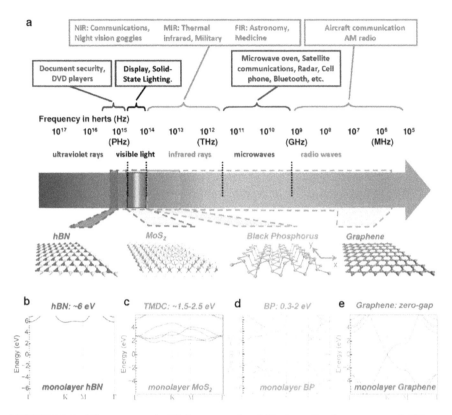

FIGURE 3.33   Electromagnetic (EM) wave and mobility/on–off ratio spectrum of different 2D material. a) The bandgap range and electromagnetic wave spectrum of different types of 2D layered nanomaterials. b–e) The various ranges of frequency conforming the bandgaps of 2D layered materials such as hBN, TMDC, BP, and graphene, and their applications in optoelectronics are also indicated. Adopted with permission from Xia et al. (2014).

3.4.3   IN-PLANE ANISOTROPY

Recently, BP has tremendous applications in photonics nanoelectronics and may well offer favorable advantages over TMDs and graphene due to its high mobility, tunable narrow bandgap, and mechanical strength, and contain new functionalities due to their porous structure. Additionally, for designing new devices and opportunities in optoelectronics, BP has unique in-plane anisotropy (2, 6) properties. Anisotropy is responsible for the most exciting application of BP that might be studied further in future. Other relatively lesser-known layered TMDs other than MoS_2, $MOSe_2$, WS_2, and WSe_2, such as Rhenium diselenide ($ReSe_2$) and Rhenium disulfide (ReS_2), have such kind of anisotropy-like properties, and together they might be applied to a new domain of photonics and electronics device research. For example, i) in the case of plasmonic devices having intrinsic anisotropy in their resonance properties; ii) in the directions of heat and electron transport, there is a high-efficiency thermoelectric using the orthogonality (Ling et al. 2015).

3.4.4 THERMAL PROPERTIES OF BACK PHOSPHORUS

Due to emerging semiconducting property and puckered structure geometry (Figure 3.34a–b) BP is considered an active area of research worldwide. Although a lot of research work has been done to explore the electrical properties of BP, there is still a deficiency of experimental studies on the thermal transport properties of BP. In reported literature, an extremely low value of thermal conductivity of around ≈10 W/ m K was established in bulk samples of BP (Slack 1965).

On the other hand, there are significant research publications (Ong et al. 2014; Zhang, Pei, et al. 2016; Qin et al. 2014, 2015; Hong et al. 2015; Liu and Chang 2015; Jain and McGaughey 2015) related to the thermal transportation proper- ties (TTP) of BP, such as its anisotropic possessions. In most of the cases, the Boltzmann transport equation (BTE) is applied for the extraction process of the intrinsic in-plane thermal conductivity in both directions such as armchair (AM) and zigzag (ZZ), which are responsible for high anisotropic ratio value of around 3.5. TEM is used for two highly symmetric directions for an exfoliated BP sample (Figure 3.34c–d), Raman technique was introduced for measuring the thermal con- ductivity value of FL-BP near the room temperature, and *ab initio* phonon disper- sion calculations were used for detail explanation (Luo et al. 2015). From the three mentioned Raman peaks such as (Ag^1, B_2g, Ag^3), it was concluded that due to high temperature sensitivity, as well as Raman intensity value, the A2g peak is selected as the thermometer/temperature detector (Figure 3.34e–f) and for BP films that are thicker than 15 nm, the measured ZZ and AM thermal conductivities values were found to be about in the range of 40 and 20 W m^{-1} K^{-1}, respectively, representing the significant anisotropy.

Another study using the micro-bridge method (Lee et al. 2015) reported similar results when BP nanoribbons were used as characteristic material. Additionally, it was found that thickness of nanoribbon and temperature have a great impact on thermal conductivity, and it was found that the BP nanoribbon having the peak value of 170 nm thickness had different thermal conductivities values of about 24 W/ mK and 21 W/ mK at temperature 100 K and 150 K, respectively. On the other hand, Jang et al. (2015) used the TDTR technique and measured the relatively thick value of exfoliated BP nanoflakes with thickness value from 138 nm to 552 nm in range. They used both kinds of TDTR technique such as beam-offset and conventional TDTR procedure and measured both the in-plane and cross-plane thermal conductivity, and strongly detected the anisotropic in-plane thermal conductivities (34 ± 4 W/ mK and 86 ± 8 W /mK) along both the AM and the ZZ ways, respectively.

Zhu et al. (2016) used first-principles calculation for 3D anisotropic values based on thermal conductivities of BP and along the direction of different crystalline orientations clarified the findings with the help of structural asymmetry-convinced assembly values and variation in velocity. The relation between thermal conduct- ivity and thickness of the film of BP is likened to other materials having 2D layers such as graphene and MoS_2. It was discovered that the out-of-plane acoustic (ZA) style contributed to the in-plane thermal transportation, and was usually applied to explain the thickness-based performance of different film of BP (Jang et al. 2015).

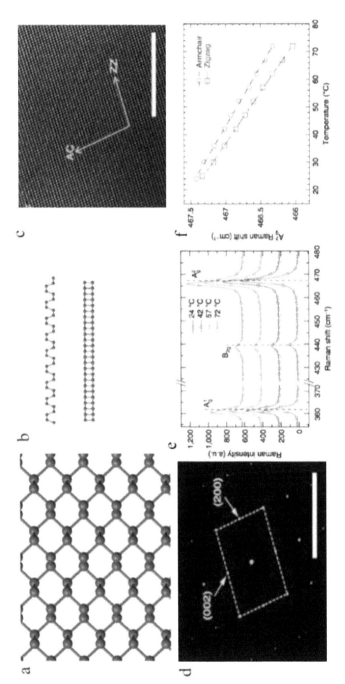

FIGURE 3.34 Structure characterization analysis of BP crystal. a) Top view puckered structure geometry analysis. b) Side looking representation of monolayer phosphorene. Adapted with permission from Wang et al. (2017), Copyright © 2017, WILEY–VCH Verlag GmbH & Co. KGaA, Weinheim. c) TEM used for two highly symmetric directions for an exfoliated BP sample nanoflake. d) SEAD pattern from c) image confirming the BP through d-spacing. Reproduced under the terms of the Creative Commons Attribution 4.0 International. Copyright © the authors, published by Springer Nature (Lee et al. 2015). e&f) Raman peaks such as (A^1g, B_2g, A^3g), it was concluded that due to high-temperature sensitivity and Raman intensity value the A2g peak, is selected as the thermometer/temperature detector. Reproduced under the terms of the Creative Commons Attribution 4.0 International. Copy right © the authors, published by Springer Nature (Luo et al. 2015).

For in-plane thermal conductivity value, ZA mode's influence is relatively less in the case of BP-like 2D layered materials having a structure with non-planar geometry. In the whole process, the surface scattering is significant and considered the leading mechanism (Jain and McGaughey 2015). On the other hand, the ZA mode dominates in graphene-like flat plane structures crushed in a few-layer form (Seol et al. 2010). Therefore, with increasing the thickness value of graphene-like layered materials, the value of in-plane thermal conductivity decreases. There is an increase in thermal conductivity value in the case of BP-like materials. Although a lot of research work has been done to explore the thermal transport properties of 2D material there is still much more to do, and several questions/problems need to be addressed, such as the shortage of techniques that might be helpful for measuring the exact monolayer structures (except for MoS_2 and graphene). Further exploration is needed into the intrinsic thermal properties of other 2D layered materials (2DLMs), and there is a need to improve the reproducibility of the developed method in several features. Reproducibility is a big challenge, and there is a deficiency of reported experimental results with the theoretical and computational analysis. In particular, there are some controversies in the relation between the thickness of layers and thermal conductivity values, and there is still a shortage of published data where the direct relation between the thermal conductivity and thickness of layers is analyzed in other 2DLMs except the experimental results in case of graphene (Eswaraiah et al. 2016).

3.4.5 MECHANICAL PROPERTIES OF BLACK PHOSPHORUS

Recently, phosphorene-based fabricated crystal structures have been found to have multidirectional applications and potential demand in electronics. The phosphorene's electronic-related characteristics (single layer) and FL-BP were significantly modified and demonstrated by the mechanical strain. 2D layered materials, such as MoS_2 and graphene, can sustain a large strain (greater than or equal to 25 percent) and, possess great mechanical flexibility (Kim et al. 2009; Lee et al. 2008; Castellanos-Gomez et al. 2012). Wei and Peng (2014) examined 2D few-layer BP and monolayer phosphorene's mechanical properties using a theoretical first principle DFT-based calculations and concluded about superior mechanical flexibility of the material. It was concluded that the SL-BP can hold a value of stress up to 8 GPa and 18 GPa in the armchair and zigzag directions, respectively. Moreover, few-layer BP can sustain strain up to around 32 percent while SL-BP can hold critical strain value up to 30 percent, and this limit of strain for phosphorene might be due to the unique crystal structure puckered type geometry (Figure 3.35a–b). It was further analyzed when there is tensile strain is applied in the armchair direction. As a result, there is no effect on the P–P bond lengths but effectively flattens the puckered structure of phosphorene. Also, the mandatory strain energy is significantly reduced due to these compromised dihedral angles. Compared to other 2D materials, there is much smaller Young's modulus in phosphorene, suggesting great applications of the materials showing the strong value of anisotropy and large-magnitude-strain engineering (Wei and Peng 2014).

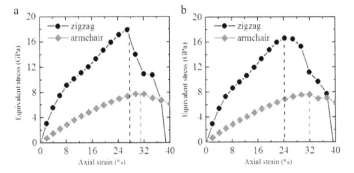

FIGURE 3.35 The strain-stress relation for different BP layers a) for single layer and b) few-layer phosphorene-based structures. The SL-BP can survive stress around 8 GPa, 18 GPa in the armchair and zigzag directions, and the equivalent critical strains value are 27 percent (in zigzag structure) and ~30% (armchair direction). The ideal strengths/critical strains value for the FL-BP are approximately around 16 GPa/24 percent and around 7.5 GPa/32 percent in zigzag and armchair direction directions. Adapted with permission from Wei and Peng (2014), Copyright © 2021. All rights reserved.

3.4.6 THE OPTICAL PROPERTY OF BLACK PHOSPHORUS

For semiconductor-based photonic devices and their applications, the presence of tunable optical-based material is an extremely anticipated property. It is noted that the value of absorption spectra in the case of multilayer BP is varying sensitively and much depends on the parameters such as the thickness of the layers, light polarization, and doping. It mainly deals with various frequencies values ranging from 2500 to 5000 cm^{-1}, which exist in the technologically relevant state at the mid- to near-infrared spectrum (Xia, Wang, and Jia 2014; Low et al. 2014).

3.4.6.1 Linear Optical Properties

For semiconducting 2D BP material, optical property's key role is extremely anticipated distinguishing for the photonic devices and anisotropic optical spectra of quasi-2D BP modified by the parameters like doping strain and thickness. There is an excellent interaction of thickness of BP layers on optical properties. A series of theoretical works published on few-layer optical properties strongly depends on BP films' thickness during the past decade.

Low et al. (2014) used the Kubo formula within a low-energy Hamiltonian to find thin films' optics-related characteristics based on multilayer BP. The apparent anisotropic optical conductivity of thin films based on a few-layer BP can be successfully reproduced using the calculation in Figure 3.36a. In the x-track above the band advantage, the oscillation of optical conductivity is possibly due to sub-band transitions. It was concluded that this optical conductivity is concerned with the interlayer coupling, the optical anisotropic, and absorption, in contrast, the value of effective anisotropic mass. As the sample thickness decreases from 20 nm to 4 nm, there is an increase in the absorption edge value from ~0.3 to ~0.6 eV while the bandgap

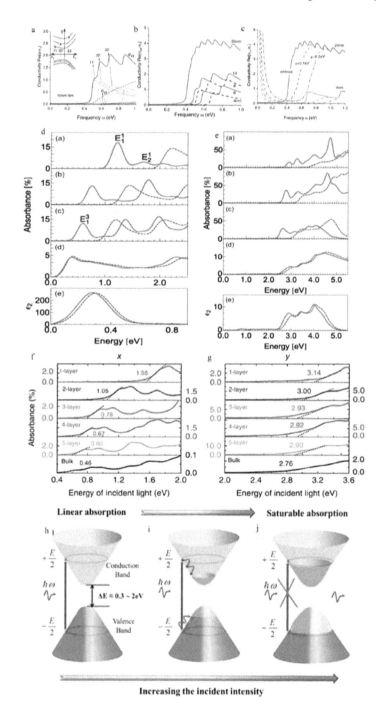

FIGURE 3.36 a) For a 10-nm-thick intrinsic layer of BP, the central portion of optical conductivities are $\text{Re}(\sigma_{xx})$ and $\text{Re}(\sigma_{yy})$, i.e., the Fermi resonance level is situated at mid-gap. Inter sub-band assistances to the σ_{xx} are diagrammed where the respective optical transition

FIGURE 3.36 Continued

processes are signified in the inset. Conductivities are regularized due to σ_0 = e2/4h. b) For intrinsic BP, the Re(σ_{xx}) of different film thicknesses of BP as indicated. c) For BP, Re(σ_{xx}) is calculated for 4- and 20-nm-thick films having different chemical potential μm. The black phosphorus absorption spectrum for various thickness and orientation. Adapted with permission from Low et al. (2014), Copyright © 2019–2020, Aptara Inc. All rights reserved. The theoretical calculation by DFT was d) and e). The figures d) and e) were reproduced with the permission of Tran et al. (2014) Copyright © 2019–2020, Aptara Inc. All rights reserved. f) and g) were the experimental result. Reproduced under the terms of the Creative Commons Attribution 4.0 International. Copyright © the authors, published by Springer Nature (Qiao et al. 2014) h–j). The schematic diagram shows the representation of the saturable absorption process in BP. Reprinted with permission from Lu et al. (2015), Copyright © The Optical Society.

does not change (Figure 3.36b). The absorption edge of BP models markedly blue shifts, possibly because of the alteration of BP's Fermi level or most probably due to increasing the density level of electrons, mentioned in Figure 3.36c. Besides, the narrow bandgap of BP can also be organized by local strain engineering, and for other 2D materials this can also be appropriate due to their unprecedented flexibility. Despite the unique puckered structure of BP, it is susceptible to the strain, unlike the other 2D counterparts such as graphene and TMDCs.

Quereda et al. discussed the optical properties of 10 nm exfoliated BP thin films as a result of strain-deformation onto pre-stretched polydimethylsiloxane (PDMS) polymer substrate (Quereda et al. 2016; Ge et al. 2015). It was analyzed that the distortion pattern is fashioned on the surface of the polymer after releasing the stretch, which might be because i) the BP sample was treated to period strain, and ii) the absorption edge transfers to redshifts (blueshifts) for compressive (tensile) strain-engineered based BP when it is compared to the bandgap value closing (opening). The total value of the absorption-based edge is extended up to ~0.7. The authors also verified that using a tight-binding model, plus or minus 5 percent strain, could convince approximately (~800 m eV) modulation of bandgap (Wang and Lan 2016). The characteristics like puckered honeycomb lattice structure and unique band structures are responsible for the polarization-based linear optical response of BP. The reflection and linear absorption spectra based on the few BP layers are shown in Figure 3.36d (Qiao et al. 2014; Tran et al. 2014). On behalf of mentioned evidence like threshold values collected from the absorption spectra, the layers' interaction on the absorption spectra is clearly shown in Figure 3.36e. The tunable bandgap range covered from ultraviolet to mid-infrared in BP is responsible for fabricating broadband-based optical devices. The strong anisotropic phenomenon from single layer to bulk is confirmed from the robust absorption process that occurs along the x-axis (armchair) and y-axis (zigzag), as shown in Figure 3.36f–g. The variation belonging to the optical response is consequently responsible for the polarization dependence (Liu, Huo, et al. 2015; Lan et al. 2016; Hong et al. 2014). Additionally, the optical response concerned with BP is responsible for approving the anisotropic behavior. The assertive anisotropic optical response behavior concerned with the BP might be responsible for the polarization-based detection process (Ribeiro et al. 2015).

3.4.6.2 Nonlinear Optical Properties

The nonlinear optical-based properties of 2D graphene, TMDs, and BP were studied in the past decade by a few researchers worldwide. During analysis narrow and direct bandgaps of BP were found, which fluctuate between single layer (2eV) to bulk (0.3 eV) with multiple layers and are positioned between graphene (zero bandgap) and TMDCs (relatively large bandgap). The features related to bandgap make them available to BP to fill the shortcoming and gap in photonics (Ling et al. 2015; Wang and Lan 2016). In 2015, after the confirmation of multilayer BP's nonlinear optical response behavior, it was widely applied in broadband laser and potential passive devices such as near-infrared and mid-infrared photonic systems (Lu et al. 2015; Wang, Huang, et al. 2015; Zheng et al. 2015; Wang et al. 2016; Liu et al. 2016; Woodward et al. 2016).

Different research groups successfully measured the nonlinear absorption behavior of BP by an open-aperture z-scan approach (Lu et al. 2015; Guo et al. 2015). This approach successfully accepts the chemically exfoliated BP suspension rather than solid-state exfoliated samples. Although there are some benefits in introducing the liquid-phase exfoliation, some shortcomings may negatively disturb the experiments. First, there are many layers present in the dispersion of BP, but only an average result for the counting of the layers is considered at the time of measurement. The number of defects found due to applying the chemical growth process may amend the concerned material's intrinsic characteristics (Zhang, Yu, et al. 2016). The schematic representation of the z-scan strategy can measure the refraction index coefficient and nonlinear absorption of a particular material at various wavelengths' range (Pálfalvi et al. 2009; Sheik-Bahae et al. 1990; Sheik-Bahae, Said, and Van Stryland 1989). The polarization P varies in the medium, on behalf of nonlinear susceptibility theory, could be mentioned as (3.1):

$$P = \varepsilon 0 \chi^{(1)} E + \varepsilon_0 \chi^{(2)} EE + \varepsilon_0 \chi^{(3)} EEE + \cdots \tag{3.1}$$

Where ε_0 shows the value of the dielectric constant value of vacuum, is the n-order value of susceptibility $\varepsilon_0 \chi^{(n)}$ tensor, E represents the value of the electric field. In the second-order term, there is a combination of sum-frequency (SF) as well as difference frequency (DF), second harmonic generation (SHG), and optical parametric amplification (OPA). Furthermore, the third-order term in equation (3.1) represents the connection to the Kerr effect, saturation absorption (SA) process, two-photon absorption (TPA), third-harmonic generation (THG), etc. Meanwhile, the imaginary part represents the triple frequency strategy, including third-harmonic generation (Mikhailov 2014; Nasari and Abrishamian 2015), nonlinear absorptions (Xin et al. 2018; Attaccalite et al. 2018; Feng et al. 2016; Cheng et al. 2019; Feng, Qin, and Liu 2018), parametric process (Wu et al. 2015; Jakubczyk et al. 2016), and Raman effect; on the other hand, its real part represents the Kerr nonlinearity phenomenon. In terms of nonlinear-based optical characteristics, the main focus is on Kerr nonlinearity and the saturable-based absorption process. Their proposed use in various fields of pulsed lasers, including solid-state lasers, waveguide lasers, fiber lasers, and related non-linear-based optical processes, are based on the above-mentioned evidence. The saturable absorption process of a material is a characteristic where light's absorption phenomenon is gradually reduced when light increases. In fact, under the influence of laser field, the value of effective refractive index n of layered based materials is expressed under laser intensity I as in equation (3.2).

$$n = n^0 + n_2 I \qquad (3.2)$$

where n^0 stands for the linear part. The Kerr nonlinearity n^2 relay on the real part of (3.2). The imaginary part of (3.2) represents the nonlinear-based absorption phenomenon of layered material with some input laser intensity. The multiple-photon absorption phenomenon and saturable absorption are two different kinds of nonlinear type absorption. It is important to note that the multi-photon absorption phenomenon becomes very significant in a very high laser intensity value (Zhang, Dong, et al. 2015; Dong et al. 2016; Zhou and Ji 2017). In the case of two-photon absorption, the saturable absorption formula can be derivative from (3.3):

$$\alpha = \alpha_{ns} + (\alpha_s / 1 + I / I_{sat}) + \beta I \qquad (3.3)$$

Where β represents the absorption process of two-photon, α_{ns} represents the unsaturated based absorption process, i.e., intensity-free damage, I designate the input laser-based intensity, S stands for the saturable based absorption and I_{sat} function as the saturable intensity. In the little value of laser intensity, the absorption value is count as $\alpha_{ns} + \alpha_s$. In contrast, in the case of a high laser intensity value, the absorption process equals $\alpha_{ns,}$ and the maximum displacement value of absorption is α_s.

The mechanistic approach related to the saturable absorption process of BP could be elaborate by the absorption process based on single-photon and two-stage-based energy-band models. Figure 3.36h–j represents the schematic process related to the excitation on nonlinear and linear alight-based absorption (Lu et al. 2015). When a beam of light with a higher photon energy E than the bandgap of BP, the process of ionization took place, and the electrons present in the region of the valence band (VB) could be moved toward the conduction band (CB) after the process of excitation, as shown in Figure 3.36h. After the process of excitation, the hot electrons thermalize to start a hot Fermi-Dirac-based circulation. After that, processing can stop the new inter-band excitation process in the VB range with the energy of $\kappa_B Te$, as represented in Figure 3.36i. As a result of high-intensity incident light, the produced electrons may cause such states where half of the photon energy is captured. As a result, it stops the other absorption process, as represented in Figure 3.36j (Chen et al. 2020; Huang, Dong, et al. 2017).

In 2015, Hanlon et al. introduced the liquid phase-based exfoliation phenomenon in the organic solvent that is N-cyclohexyl-2-pyrrolidone (CHP) solvent successful exfoliation process of high-quality few-layer BP nanolayers on a large scale (Hanlon et al. 2015).

Additionally, they also presented for the first time the nonlinear broadband ultrafast optical effect of BP. They concluded that the value of transmittance is gradually enhanced as a result of increasing the light intensity. Moreover, BP has a reliable saturable absorption effect rather than graphene. These findings suggest that BP has a wide range of potential for practical applications such as saturable absorbers, optical switches to mode-locking, and Q-switching devices (Chen et al. 2020). It is found that the morphology of BP has a significant impact on the optical properties. In the case of the nominal size of BP in a few nanometers (nm), the quantum effect can lead to nonlinear optical characteristics.

In another study, Xu et al. successfully used the solvothermal-based method to prepare BP quantum dots (BPQDs) with an average particle size of 2.1± 0.9 nm (Xu et al. 2016). It is found that the nonlinear saturable absorption efficiency of the BPQDs is more efficient than the BP nanosheets. Consequently, the modulation depth value is found as 36 percent, and saturable intensity is about 3.3 GW/cm^2. The BPQDs was then applied as a saturable absorber in the passively mode-locked fiber as part of ultrafast laser generation (Xu et al. 2016). Besides saturable absorption characteristics, BP also associated the optical Kerr effect, which alters the susceptibility value due to response in the electric field's value (Xu et al. 2016). Different factors such as cross-phase modulation (XPM), phase modulation (SPM), and phenomena based on self-focusing/defocusing are associated with the optical Kerr effect. Zheng et al. measured the value of the third-order nonlinear based refractive index of BP by z-scan based measurement process under 800 nm femtosecond pulsed laser-based excitation process, which is found as ~(6.8± 0.2)×10–13m^2/W (Zheng et al. 2015).

In another study Zhang et al. analyzed the nonlinear refractive index characteristics of the BP nanoflake based suspensions in the range of 350–1160 nm with spatial self-phase modulation (SSPM) (Zhang, Yu, et al. 2016). The remarkable nonlinear optical characteristics associated with the BP make it a capable applicant as a nonlinear based optical resource for further academic and industrial uses (Zhang, Yu, et al. 2016). After the overall discussion, the linear and nonlinear ultrafast experimental approach's main dissimilarities are the earlier emphases on the degenerate process and nonlinear-based optical regime. At the same time, the latter is associated with the recombination process and hot carrier cooling.

3.4.7 Electrical Properties of Black Phosphorus

2D nanomaterial is an emerging class of materials for nanoelectronics and photonics. For practical application (from the academic lab to industry), there is a demand for candidate systems with a sufficiently large electronic bandgap and high carrier mobility required compared to graphene.

Akahama, Endo, and Narita (1983) measured the electrical conductivity of BP as a function of pressure and temperature up to 8000 kg/cm^2 and 350°C. Additionally, as a function of temperature at atmospheric pressure, the Hall constant of the same material has been measured. It was analyzed that at high temperatures, the phosphorus is an intrinsic semiconductor with a gap width of 0.33 eV and at low temperatures, p-type impurity-based conduction is detected. Moreover, the mobilities 220 cm^2/volt sec and at 27°C are 350 cms/volt sec for the electrons and holes, respectively. In another study, Qiao et al. (2014) showed theoretically that the layer thickness of 2D semiconductor controls the value of direct bandgap, high transport anisotropy, and high carrier mobility. It was also found that the narrow bandgap value is decreased for a monolayer ranging from 1.51 eV to 0.59 eV for FL-BP. BP is distinguished from layered materials of other 2D structures such as silicone (Vogt et al. 2012; Houssa et al. 2011), h-BN5 (Liu, Feng, and Shen 2003), and germanene (Houssa et al. 2011) due to the anisotropy of optical and electric properties. Additionally, the mobility of the electrons are usually less than holes and showing the value of higher conductivity usually found the focus, which is vertical to the troughs (the x-way) and the mobility value in the x-track gradually enhanced from a monolayer (600 cm^2/Vs) to over for FL-BP (4,000 cm^2/Vs) of BP. It was stated that a single layer of BP is an active area of research and place as a special 2D material for optoelectronic devices due to the expected high hole mobility value in the range of 10,000–26,000 cm^2/Vs (Qiao et al. 2014; Akahama, Endo, and Narita 1983). These mobility results recommended a much stronger coupling between the layer found in FL-BP than in TMDCs or graphene. The electronic structures of FL-BP, such as the view from the top side of the single-layer structure and the related Brillouin zone and views from the side of the atomic structure of the bilayer structures, are shown in Figure 3.37a–b and Figure 3.37c–d respectively. These results strongly recommended that the FL-BP provides a tremendous and reliable candidate for future applications in optoelectronics and electronics devices (Qiao et al. 2014).

3.4.8 BIODEGRADABLE PROPERTY OF BLACK PHOSPHORUS

Biocompatibility is an imperative essential for BP based nanomaterials used in the field of biomedicine. BP is attractive as an inorganic nano-agent because the phosphorus (P) is working as a benevolent element constructing up approximately 1 percent bodyweight of the total basic bone component in the human physique, as well as due to its inherent biocompatibility (Comber et al. 2013; Childers et al. 2011; Pravst 2011).

Additionally, phosphate and phosphatase are final degradation products that are nontoxic (Ling et al. 2015), and both of them are accepted by the human body (Childers et al. 2011; Pravst 2011). That's why BPQDs with biocompatibility and good photothermal efficiency are potential therapeutic agents.

In the presence of phosphate-buffered saline (PBS; pH 7.4), the biodegradation behavior of the bare/simple BPQDs and PLGA functionalized BPQDs composite NSs is studied using a horizontal type shaker at a temperature of 37°C by Chu et al. Due to its appropriate biodegradation behavior in vitro (Liu, Song, et al. 2015), the simple BPQDs with a combination of PBS are subjected to a shaker for faster degradation; as a result, there is a greater absorption as well as photothermal loss during 24 hours when compared with natural dispersion in water. Despite this, a combination of BPQDs/PLGA NSs, during the initial 24 hours, maintains stability and performance (Figure 3.38a–b). Additionally, PLGA functionalized BPQDs composite NSs in PBS were further analyzed for the long-term biodegradability after eight weeks and the photothermal and absorption presentation of the composite PLGA functionalized

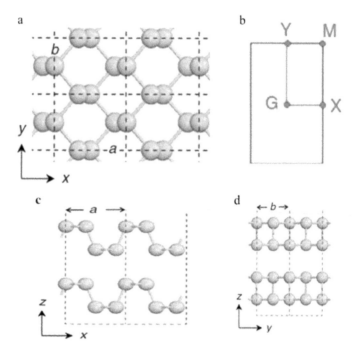

FIGURE 3.37 Electronic structures representation of FL-BP. a&b) The electronic structures of FL-BP such as the view from the top side of the single-layer atomic structure of BP. c&d) The related Brillouin zone and views from the side of bilayer side of atomic structure. Reproduced under the terms of the Creative Commons Attribution 4.0 International. Copyright © the authors, published by Springer Nature (Qiao et al. 2014).

BPQDs composite NSs. Figure 3.38c shows the remaining/residual masses of the PLGA functionalized BPQDs composite NSs during the process of degradation and the first week of analysis the residual weight reveals a downward trend having small degradation rate but it gradually accelerates during the passage of time and represents around 80 percent loss after the passage of eight weeks.

Figure 3.38d shows the surface and internal morphology (size and shape) variation after the initial stage and with time. It is clearly observed from morphology that after one week, the NSs maintain integrity. Additionally, after the passage of four weeks and eight weeks, the process of degradation is visible and it can be best analyzed as a result of variation in morphology changes. It is clearly observed that after the passage of time duration of around eight weeks, the size and shape of the nanosheets (NSs) are fully vanished and get some disordered form and very small contents of residues having BPQDs can be seen (indicated by arrow) (Shao et al. 2016).

Figure 3.38e illustrates the mechanistic approach of the degradation process. The process of hydrolysis occurs; thus, the external PLGA shells degrade gradually when the physiological environment is provided to BPQDs/PLGA NSs (Anderson and Shive 1997; Avgoustakis et al. 2002). Moreover, as a result of the degradation of PLGA, it causes the discharge of the interior BPQDs, which is degraded promptly and disrupts

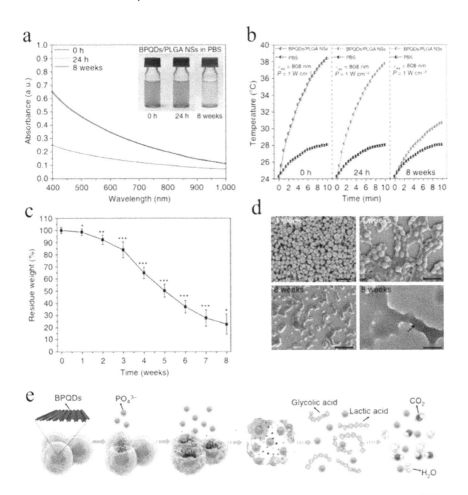

FIGURE 3.38 Biodegradation performance of black phosphorus quantum dot (BPQDs). a) UV visible light spectra of the composite PLGA functionalized BPQDs NSs (with 10 ppm BPQDs) is dispersed in PBS for different time intervals (0 hour, 24 hours, and after eight weeks) and the analogous snapshot are shown in the inset. b) The photothermal temperature graph of the composite PLGA functionalized BPQDs NSs after dispersion in PBS for different time intervals such as 0 hour, 24 hours, and after the time duration of eight weeks, and 808 nm laser was used for irradiation purpose for ten minutes. c) The remaining/residual masses of the BPQDs/PLGA composite NSs, during the process of degradation and the first week of analysis the residual weight reveals a descending trend having a small rate of degradation but it gradually accelerates with time and represents around 80 percent loss after the passage of eight weeks. d) After the process of degradation study in PBS for time duration of one, four, and eight weeks, the SEM images (scale bars, 500 nm) of the BPQDs/PLGA NSs and together with the equivalent TEM image after degradation for eight weeks (scale bar, 200 nm) of the NSs. e) In case of the physiological environment there is a process of degradation study of the composite of PLGA functionalized BPQDs NSs. Reproduced under the terms of the Creative Commons Attribution 4.0 International. Copyright © the authors, published by Springer Nature (Shao et al. 2016).

rapidly when there is no protection with PLGA. Nontoxic phosphate (Ling et al. 2015) and phosphate are the final products obtained due to the degradation of the BPQDs, both of which are commonly found as part of the human body (Childers et al. 2011; Pravst 2011). In case of *in vivo* treatment, the unique biodegradable behavior of the PLGA functionalized BPQDs composite NSs facilitates the inoffensive authorization from the body with a specific time duration after accomplishing their therapeutic functions but avoids the rapid degradation as a result of the optical treatment.

In another study Qiu et al. (2018) investigated the cytotoxicity of BPNSs, after conducting a brief analysis report on BPNSs interaction with four different types of cells such as human breast cancer cells (MDA-MB-231), human lung carcinoma cells (A549), human cervical cancer cell (HeLa), and mouse melanoma cells (B16) for the time duration of 48 hours. Cell cytotoxicity assays confirmed that the absence of cytotoxicity for all the selected cells even after injecting the high dose of concentration (200 μg/mL). This evidence strongly recommended the potential biosafety of BPNSs (Qiu et al. 2018).

Finally, such a unique set of properties and stable BP based nanostructures (as a result of prepared composite with polymers and heterostructures TMDCs, graphene, etc., with improved properties) might be applicable for energy storage (Wu, Hui, and Hui 2018; Qiu et al. 2017), electronics (Hong et al. 2014; Ling et al. 2015), catalysis (Wu et al. 2018; Jiang et al. 2016) and biomedical fields (Yang et al. 2018), and sensors (Lee et al. 2018), etc.

3.5 SUMMARY

Different synthetic techniques, such as top-down and bottom-up approaches, as well as BP properties, are discussed in this chapter. Different structures of 2D BP nanomaterials—including quantum dots, nanotubes, bulk, and monolayers, besides other characteristics such as semi-conductive values, chemical and thermal stability—have remarkable optical and biocompatible, electrical and mechanical characteristics. It is highly desired to address problems related to stability issue of BP, synthetic strategies, such as control morphology and high returns, for practical applications of BP-based nanostructures in electronics, photonics, biomedicine, sensing, and environmental applications.

3.6 ACKNOWLEDGMENTS

The research was partially supported by financial support from the Science and Technology Development Fund (Nos. 007/2017/A1 and 132/2017/A3), Macao Special Administration Region (SAR), China, and National Natural Science Fund (Grant Nos. 61875138, 61435010, 61805147, and 6181101252), and Science and Technology Innovation Commission of Shenzhen (KQTD20150324162703, JCYJ20150625103619275,   JCYJ20180305125141661,   and   JCYJ20170 811093453105). The authors also acknowledge the support from the Instrumental Analysis Center of Shenzhen University (Xili Campus).

REFERENCES

Abbasi, Nasir M., Haojie Yu, Li Wang, Zain-ul-Abdin, Wael A. Amer, Muhammad Akram, Hamad Khalid, Yongsheng Chen, Muhammad Saleem, and Ruoli Sun. 2015. "Preparation of silver nanowires and their application in conducting polymer nanocomposites." *Materials Chemistry and Physics* 166: 1–15.

Abellán, Gonzalo, Stefan Wild, Vicent Lloret, Nils Scheuschner, Roland Gillen, Udo Mundloch, Janina Maultzsch, Maria Varela, Frank Hauke, and Andreas Hirsch. 2017. "Fundamental insights into the degradation and stabilization of thin layer black phosphorus." *Journal of the American Chemical Society* 139(30): 10432–10440.

Ahmed, Taimur, Sivacarendran Balendhran, Md Nurul Karim, Edwin L. H. Mayes, Matthew R. Field, Rajesh Ramanathan, Mandeep Singh, Vipul Bansal, Sharath Sriram, and Madhu Bhaskaran. 2017. "Degradation of black phosphorus is contingent on UV–blue light exposure." *npj 2D Materials and Applications* 1(1): 18.

Akahama, Yuichi, Shoichi Endo, and Shin-ichiro Narita. 1983. "Electrical properties of black phosphorus single crystals." *Journal of the Physical Society of Japan* 52(6): 2148–2155.

Ambrosi, Adriano, and Martin Pumera. 2016. "Electrochemically exfoliated graphene and graphene oxide for energy storage and electrochemistry applications." *Chemistry: A European Journal* 22(1): 153–159.

Ambrosi, Adriano, Zdeněk Sofer, Jan Luxa, and Martin Pumera. 2016. "Exfoliation of layered topological insulators Bi2Se3 and Bi2Te3 via electrochemistry." *ACS nano* 10(12): 11442–11448.

Ambrosi, Adriano, Zdeněk Sofer, and Martin Pumera. 2017. "Electrochemical exfoliation of layered black phosphorus into phosphorene." *Angewandte Chemie International Edition* 56(35): 10443–10445.

Anderson, James M., and Matthew S. Shive. 1997. "Biodegradation and biocompatibility of PLA and PLGA microspheres." *Advanced Drug Delivery Reviews* 28(1): 5–24.

Attaccalite, Claudio, Myrta Grüning, Hakim Amara, Sylvain Latil, and François Ducastelle. 2018. "Two-photon absorption in two-dimensional materials: The case of hexagonal boron nitride." *Physical Review B* 98(16): 165126.

Avgoustakis, K., A. Beletsi, Z. Panagi, P. Klepetsanis, A. G. Karydas, and D. S. Ithakissios. 2002. "PLGA–mPEG nanoparticles of cisplatin: in vitro nanoparticle degradation, in vitro drug release and in vivo drug residence in blood properties." *Journal of Controlled Release* 79(1–3): 123–135.

Bao, Ta Na, Ojin Tegus, Hasichaolu, Ning Jun, and Narengerile. 2018. "Preparation of black phosphorus by the mechanical ball milling method and its characterization." *Solid State Phenomena* 271: 18–22.

Batmunkh, Munkhbayar, Munkhjargal Bat-Erdene, and Joseph G. Shapter. 2016. "Phosphorene and phosphorene-based materials: prospects for future applications." *Advanced Materials* 28(39): 8586–8617.

Batmunkh, Munkhbayar, Munkhjargal Bat-Erdene, and Joseph G. Shapter. 2018. "Black phosphorus: synthesis and application for solar cells." *Advanced Energy Materials* 8(5): 1701832.

Brent, Jack R., Nicky Savjani, Edward A. Lewis, Sarah J. Haigh, David J. Lewis, and Paul O'Brien. 2014. "Production of few-layer phosphorene by liquid exfoliation of black phosphorus." *Chemical Communications* 50(87): 13338–13341.

Brent, Jack R., Ashok K. Ganguli, Vinod Kumar, David J. Lewis, Paul D. McNaughter, Paul O'Brien, Priyanka Sabherwal, and Aleksander A. Tedstone. 2016. "On the stability of surfactant-stabilised few-layer black phosphorus in aqueous media." *RSC Advances*

6(90): 86955–86958.Bullett, D. W. 1985. "Valence-band structures of phosphorus allotropes II." *Solid State Communications* 55(3): 257–260.

Cai, Jinming, Pascal Ruffieux, Rached Jaafar, Marco Bieri, Thomas Braun, Stephan Blankenburg, Matthias Muoth, Ari P. Seitsonen, Moussa Saleh, Xinliang Feng, Klaus Müllen, and Roman Fasel. 2010. "Atomically precise bottom-up fabrication of graphene nanoribbons." *Nature* 466: 470.

Cai, Yongqing, Qingqing Ke, Gang Zhang, Yuan Ping Feng, Vivek B. Shenoy, and Yong-Wei Zhang. 2015. "Giant phononic anisotropy and unusual anharmonicity of phosphorene: interlayer coupling and strain engineering." *Advanced Functional Materials* 25(15): 2230–2236.

Carvalho, Alexandra, Min Wang, Xi Zhu, Aleksandr S. Rodin, Haibin Su, and Antonio H. Castro Neto. 2016. "Phosphorene: from theory to applications." *Nature Reviews Materials* 1(11): 16061.

Castellanos-Gomez, Andres, Menno Poot, Gary A. Steele, Herre S. J. Van der Zant, Nicolás Agraït, and Gabino Rubio-Bollinger. 2012. "Mechanical properties of freely suspended semiconducting graphene-like layers based on MoS 2." *Nanoscale Research Letters* 7(1): 233.

Castellanos-Gomez, Andres, Leonardo Vicarelli, Elsa Prada, Joshua O. Island, K. L. Narasimha-Acharya, Sofya I. Blanter, Dirk J. Groenendijk, Michele Buscema, Gary A. Steele, and J. V. Alvarez. 2014. "Isolation and characterization of few-layer black phosphorus." *2D Materials* 1(2): 025001.

Caswell, K. K., Christopher M. Bender, and Catherine J. Murphy. 2003. "Seedless, surfactantless wet chemical synthesis of silver nanowires." *Nano Letters* 3(5): 667–669.

Chen, Long, Guangmin Zhou, Zhibo Liu, Xiaomeng Ma, Jing Chen, Zhiyong Zhang, Xiuliang Ma, Feng Li, Hui-Ming Cheng, and Wencai Ren. 2016. "Scalable clean exfoliation of high-quality few-layer black phosphorus for a flexible lithium ion battery." *Advanced Materials* 28(3): 510–517.

Chen, Xing, Joice Sophia Ponraj, Dianyuan Fan, and Han Zhang. 2020. "An overview of the optical properties and applications of black phosphorus." *Nanoscale* 12(6): 3513–3534.

Chen, Yu, Guobao Jiang, Shuqing Chen, Zhinan Guo, Xuefeng Yu, Chujun Zhao, Han Zhang, Qiaoliang Bao, Shuangchun Wen, and Dingyuan Tang. 2015. "Mechanically exfoliated black phosphorus as a new saturable absorber for both Q-switching and mode-locking laser operation." *Optics Express* 23(10): 12823–12833.

Cheng, Jingxin, Di Huang, Tao Jiang, Yuwei Shan, Yingguo Li, Shiwei Wu, and Wei-Tao Liu. 2019. "Chiral selection rules for multi-photon processes in two-dimensional honeycomb materials." *Optics Letters* 44(9): 2141–2144.

Childers, Daniel L., Jessica Corman, Mark Edwards, and James J. Elser. 2011. "Sustainability challenges of phosphorus and food: solutions from closing the human phosphorus cycle." *Bioscience* 61(2): 117–124.

Coleman, Jonathan N., Mustafa Lotya, Arlene O'Neill, Shane D. Bergin, Paul J. King, Umar Khan, Karen Young, Alexandre Gaucher, Sukanta De, and Ronan J. Smith. 2011. "Two-dimensional nanosheets produced by liquid exfoliation of layered materials." *Science* 331(6017): 568–571.

Comber, Sean, Michael Gardner, Karyn Georges, David Blackwood, and Daniel Gilmour. 2013. "Domestic source of phosphorus to sewage treatment works." *Environmental Technology* 34(10): 1349–1358.

Dai, Jun, and Xiao Cheng Zeng. 2014. "Bilayer phosphorene: effect of stacking order on bandgap and its potential applications in thin-film solar cells." *The Journal of Physical Chemistry Letters* 5(7): 1289–1293.

Das, Saptarshi, Wei Zhang, Marcel Demarteau, Axel Hoffmann, Madan Dubey, and Andreas Roelofs. 2014. "Tunable transport gap in phosphorene." *Nano Letters* 14(10): 5733–5739.

Dhanabalan, Sathish Chander, Joice Sophia Ponraj, Zhinan Guo, Shaojuan Li, Qiaoliang Bao, and Han Zhang. 2017. "Emerging trends in phosphorene fabrication towards next generation devices." *Advanced Science* 4(6): 1600305.

Dong, Ningning, Yuanxin Li, Saifeng Zhang, Niall McEvoy, Xiaoyan Zhang, Yun Cui, Long Zhang, Georg S Duesberg, and Jun Wang. 2016. "Dispersion of nonlinear refractive index in layered WS 2 and WSe 2 semiconductor films induced by two-photon absorption." *Optics Letters* 41(17): 3936–3939.

Erande, Manisha B., Sachin R. Suryawanshi, Mahendra A. More, and Dattatray J. Late. 2015. "Electrochemically exfoliated black phosphorus nanosheets: prospective field emitters." *European Journal of Inorganic Chemistry* 2015(19): 3102–3107.

Eswaraiah, Varrla, Qingsheng Zeng, Yi Long, and Zheng Liu. 2016. "Black phosphorus nanosheets: synthesis, characterization and applications." *Small* 12(26): 3480–3502.

Fan, Shuangqing, Haicheng Hei, Chunhua An, Wei Pang, Daihua Zhang, Xiaodong Hu, Sen Wu, and Jing Liu. 2017. "Rapid thermal thinning of black phosphorus." *Journal of Materials Chemistry C* 5(40): 10638–10644.

Fan, Shuangqing, JingSi Qiao, Jiawei Lai, Haicheng Hei, Zhihong Feng, Qiankun Zhang, Daihua Zhang, Sen Wu, Xiaodong Hu, Dong Sun, Wei Ji, and Jing Liu. 2019. "Wet chemical method for black phosphorus thinning and passivation." *ACS Applied Materials & Interfaces* 11(9): 9213–9222.

Favron, Alexandre, Etienne Gaufrès, Frédéric Fossard, Anne-Laurence Phaneuf-L'Heureux, Nathalie Y. W. Tang, Pierre L. Lévesque, Annick Loiseau, Richard Leonelli, Sébastien Francoeur, and Richard Martel. 2015. "Photooxidation and quantum confinement effects in exfoliated black phosphorus." *Nature Materials* 14(8): 826–832.

Feng, Xiaobo, Xin Li, Zhisong Li, and Yingkai Liu. 2016. "Size-dependent two-photon absorption in circular graphene quantum dots." *Optics Express* 24(3): 2877–2884.

Feng, Xiaobo, Yonggang Qin, and Yu Liu. 2018. "Size and edge dependence of two-photon absorption in rectangular graphene quantum dots." *Optics Express* 26(6): 7132–7139.

Feng, Yanqing, Jian Zhou, Yongping Du, Feng Miao, Chun-Gang Duan, Baigeng Wang, and Xiangang Wan. 2015. "Raman spectra of few-layer phosphorene studied from first-principles calculations." *Journal of Physics: Condensed Matter* 27(18): 185302.

Gan, Z. X., L. L. Sun, X. L. Wu, M. Meng, J. C. Shen, and Paul K. Chu. 2015. "Tunable photoluminescence from sheet-like black phosphorus crystal by electrochemical oxidation." *Applied Physics Letters* 107(2): 021901.

Ge, Shaofeng, Chaokai Li, Zhiming Zhang, Chenglong Zhang, Yudao Zhang, Jun Qiu, Qinsheng Wang, Junku Liu, Shuang Jia, and Ji Feng. 2015. "Dynamical evolution of anisotropic response in black phosphorus under ultrafast photoexcitation." *Nano Letters* 15(7): 4650–4656.

Geim, A. K., and K. S. Novoselov. 2007. "The rise of graphene." *Nature Materials* 6: 183.

Green, Alexander A., and Mark C. Hersam. 2009. "Solution phase production of graphene with controlled thickness via density differentiation." *Nano Letters* 9(12): 4031–4036.

Guan, Liao, Boran Xing, Xinyue Niu, Dan Wang, Ying Yu, Shucheng Zhang, Xiaoyuan Yan, Yewu Wang, and Jian Sha. 2018. "Metal-assisted exfoliation of few-layer black phosphorus with high yield." *Chemical Communications* 54(6): 595–598.

Guo, Zhinan, Han Zhang, Shunbin Lu, Zhiteng Wang, Siying Tang, Jundong Shao, Zhengbo Sun, Hanhan Xie, Huaiyu Wang, Xue-Feng Yu, and Paul K. Chu. 2015. "From black phosphorus to phosphorene: basic solvent exfoliation, evolution of Raman scattering, and applications to ultrafast photonics." *Advanced Functional Materials* 25(45): 6996–7002.

Gusmao, Rui, Zdenek Sofer, and Martin Pumera. 2017. "Black phosphorus rediscovered: from bulk material to monolayers." *Angewandte Chemie International Edition* 56(28): 8052–8072.

Hanlon, Damien, Claudia Backes, Evie Doherty, Clotilde S. Cucinotta, Nina C. Berner, Conor Boland, Kangho Lee, Andrew Harvey, Peter Lynch, and Zahra Gholamvand. 2015. "Liquid exfoliation of solvent-stabilized few-layer black phosphorus for applications beyond electronics." *Nature Communications* 6(1): 1–11.

Hansen, Charles M. 2007. *Hansen solubility parameters: a user's handbook*. London: CRC Press.

Hao, Jianhua, Weidong Si, X. X. Xi, Ruyan Guo, A. S. Bhalla, and L. E. Cross. 2000. "Dielectric properties of pulsed-laser-deposited calcium titanate thin films." *Applied Physics Letters* 76(21): 3100–3102.

Hong, Tu, Bhim Chamlagain, Wenzhi Lin, Hsun-Jen Chuang, Minghu Pan, Zhixian Zhou, and Ya-Qiong Xu. 2014. "Polarized photocurrent response in black phosphorus field-effect transistors." *Nanoscale* 6(15): 8978–8983.

Hong, Yang, Jingchao Zhang, Xiaopeng Huang, and Xiao Cheng Zeng. 2015. "Thermal conductivity of a two-dimensional phosphorene sheet: a comparative study with graphene." *Nanoscale* 7(44): 18716–18724.

Houssa, M., E. Scalise, K. Sankaran, G. Pourtois, V. V. Afanas'ev, and A. Stesmans. 2011. "Electronic properties of hydrogenated silicene and germanene." *Applied Physics Letters* 98(22): 223107.

Huang, Jiawei, Ningning Dong, Saifeng Zhang, Zhenyu Sun, Wanhong Zhang, and Jun Wang. 2017. "Nonlinear absorption induced transparency and optical limiting of black phosphorus nanosheets." *ACS Photonics* 4(12): 3063–3070.

Huang, Yuan, Jingsi Qiao, Kai He, Stoyan Bliznakov, Eli Sutter, Xianjue Chen, Da Luo, Fanke Meng, Dong Su, and Jeremy Decker. 2016. "Interaction of black phosphorus with oxygen and water." *Chemistry of Materials* 28(22): 8330–8339.

Huang, Zhaodong, Hongshuai Hou, Yan Zhang, Chao Wang, Xiaoqing Qiu, and Xiaobo Ji. 2017. "Layer-tunable phosphorene modulated by the cation insertion rate as a sodium-storage anode." *Advanced Materials* 29(34): 1702372.

Hultgren, Ralph, N. S. Gingrich, and B. E. Warren. 1935. "The atomic distribution in red and black phosphorus and the crystal structure of black phosphorus." *The Journal of Chemical Physics* 3(6): 351–355.

Jain, Ankit, and Alan J. H. McGaughey. 2015. "Strongly anisotropic in-plane thermal transport in single-layer black phosphorene." *Scientific Reports* 5: 8501.

Jakubczyk, Tomasz, Valentin Delmonte, Maciej Koperski, Karol Nogajewski, Clément Faugeras, Wolfgang Langbein, Marek Potemski, and Jacek Kasprzak. 2016. "Radiatively limited dephasing and exciton dynamics in MoSe2 monolayers revealed with four-wave mixing microscopy." *Nano Letters* 16(9): 5333–5339.

Jana, Nikhil, Latha Gearheart, and Catherine Murphy. 2001. "Wet chemical synthesis of silver nanorods and nanowires of controllable aspect ratio." *Chemical Communications* 7: 617–618.

Jang, Hyejin, Joshua D. Wood, Christopher R. Ryder, Mark C. Hersam, and David G. Cahill. 2015. "Anisotropic thermal conductivity of exfoliated black phosphorus." *Advanced Materials* 27(48): 8017–8022.

Jia, Jingyuan, Sung Kyu Jang, Shen Lai, Jiao Xu, Young Jin Choi, Jin-Hong Park, and Sungjoo Lee. 2015. "Plasma-treated thickness-controlled two-dimensional black phosphorus and its electronic transport properties." *ACS Nano* 9(9): 8729–8736.

Jiang, Qianqian, Lei Xu, Ning Chen, Han Zhang, Liming Dai, and Shuangyin Wang. 2016. "Facile synthesis of black phosphorus: an efficient electrocatalyst for the oxygen evolving reaction." *Angewandte Chemie* 128(44): 14053–14057.

Kang, Joohoon, Jung-Woo T. Seo, Diego Alducin, Arturo Ponce, Miguel Jose Yacaman, and Mark C. Hersam. 2014. "Thickness sorting of two-dimensional transition metal dichalcogenides via copolymer-assisted density gradient ultracentrifugation." *Nature Communications* 5: 5478.

Kang, Joohoon, Joshua D. Wood, Spencer A. Wells, Jae-Hyeok Lee, Xiaolong Liu, Kan-Sheng Chen, and Mark C. Hersam. 2015. "Solvent exfoliation of electronic-grade, two-dimensional black phosphorus." *ACS Nano* 9(4): 3596–3604.

Kang, Joohoon, Spencer A. Wells, Joshua D. Wood, Jae-Hyeok Lee, Xiaolong Liu, Christopher R. Ryder, Jian Zhu, Jeffrey R. Guest, Chad A. Husko, and Mark C. Hersam. 2016. "Stable aqueous dispersions of optically and electronically active phosphorene." *Proceedings of the National Academy of Sciences* 113(42): 11688–11693.

Khandelwal, Apratim, Karthick Mani, Manohar Harsha Karigerasi, and Indranil Lahiri. 2017. "Phosphorene—the two-dimensional black phosphorous: properties, synthesis and applications." *Materials Science and Engineering: B* 221: 17–34.

Kim, Keun Soo, Yue Zhao, Houk Jang, Sang Yoon Lee, Jong Min Kim, Kwang S. Kim, Jong-Hyun Ahn, Philip Kim, Jae-Young Choi, and Byung Hee Hong. 2009. "Large-scale pattern growth of graphene films for stretchable transparent electrodes." *Nature* 457(7230): 706.

Kumar, Vinod, Jack R. Brent, Munish Shorie, Harmanjit Kaur, Gaganpreet Chadha, Andrew G. Thomas, Edward A. Lewis, Aidan P. Rooney, Lan Nguyen, Xiang Li Zhong, M. Grace Burke, Sarah J. Haigh, Alex Walton, Paul D. McNaughter, Aleksander A. Tedstone, Nicky Savjani, Christopher A. Muryn, Paul O'Brien, Ashok K. Ganguli, David J. Lewis, and Priyanka Sabherwal. 2016. "Nanostructured aptamer-functionalized black phosphorus sensing platform for label-free detection of myoglobin, a cardiovascular disease biomarker." *ACS Applied Materials & Interfaces* 8(35): 22860–22868.

Lan, Shoufeng, Sean Rodrigues, Lei Kang, and Wenshan Cai. 2016. "Visualizing optical phase anisotropy in black phosphorus." *Acs Photonics* 3(7): 1176–1181.

Lange, Stefan, Peer Schmidt, and Tom Nilges. 2007. "Au3SnP7@Black Phosphorus: an easy access to black phosphorus." *Inorganic Chemistry* 46(10): 4028–4035.

Late, Dattatray J. 2015. "Temperature dependent phonon shifts in few-layer black phosphorus." *ACS Applied Materials & Interfaces* 7(10): 5857–5862.

Late, Dattatray J. 2016. "Liquid exfoliation of black phosphorus nanosheets and its application as humidity sensor." *Microporous and Mesoporous Materials* 225: 494–503.

Lee, Changgu, Xiaoding Wei, Jeffrey W. Kysar, and James Hone. 2008. "Measurement of the elastic properties and intrinsic strength of monolayer graphene." *Science* 321(5887): 385–388.

Lee, Changgu, Hugen Yan, Louis E Brus, Tony F Heinz, James Hone, and Sunmin Ryu. 2010. "Anomalous lattice vibrations of single- and few-layer MoS2." *ACS Nano* 4(5): 2695–2700.

Lee, Hyun Uk, So Young Park, Soon Chang Lee, Saehae Choi, Soonjoo Seo, Hyeran Kim, Jonghan Won, Kyuseok Choi, Kyoung Suk Kang, and Hyun Gyu Park. 2016. "Black phosphorus (BP) nanodots for potential biomedical applications." *Small* 12(2): 214–219.

Lee, Kyoung, Kyung Kim, Hyeonseok Yoon, and Hyungwoo Kim. 2018. "Chemical design of functional polymer structures for biosensors: from nanoscale to macroscale." *Polymers* 10(5): 551.

Lee, Sangwook, Fan Yang, Joonki Suh, Sijie Yang, Yeonbae Lee, Guo Li, Hwan Sung Choe, Aslihan Suslu, Yabin Chen, and Changhyun Ko. 2015. "Anisotropic in-plane thermal conductivity of black phosphorus nanoribbons at temperatures higher than 100 K." *Nature Communications* 6: 8573.

Li, Likai, Yijun Yu, Guo Jun Ye, Qingqin Ge, Xuedong Ou, Hua Wu, Donglai Feng, Xian Hui Chen, and Yuanbo Zhang. 2014. "Black phosphorus field-effect transistors." *Nature Nanotechnology* 9(5): 372–377.

Li, Qiang, Qionghua Zhou, Xianghong Niu, Yinghe Zhao, Qian Chen, and Jinlan Wang. 2016. "Covalent functionalization of black phosphorus from first-principles." *The Journal of Physical Chemistry Letters* 7(22): 4540–4546.

Li, Xuesong, Weiwei Cai, Jinho An, Seyoung Kim, Junghyo Nah, Dongxing Yang, Richard Piner, Aruna Velamakanni, Inhwa Jung, Emanuel Tutuc, Sanjay K. Banerjee, Luigi Colombo, and Rodney S. Ruoff. 2009. "Large-area synthesis of high-quality and uniform graphene films on copper foils." *Science* 324(5932): 1312.

Li, Xuesong, Bingchen Deng, Xiaomu Wang, Sizhe Chen, Michelle Vaisman, Shun-ichiro Karato, Grace Pan, Minjoo Larry Lee, Judy Cha, and Han Wang. 2015. "Synthesis of thin-film black phosphorus on a flexible substrate." *2D Materials* 2(3): 031002.

Ling, Xi, Han Wang, Shengxi Huang, Fengnian Xia, and Mildred S. Dresselhaus. 2015. "The renaissance of black phosphorus." *Proceedings of the National Academy of Sciences* 112(15): 4523–4530.

Liu, Bin, Yu-wei Song, Li Jin, Zhi-jian Wang, De-yong Pu, Shao-qiang Lin, Chan Zhou, Hua-jian You, Yan Ma, and Jin-min Li. 2015. "Silk structure and degradation." *Colloids and Surfaces B: Biointerfaces* 131: 122–128.

Liu, Dahui, Bing Gu, Boxiao Ren, Changgui Lu, Jun He, Qiwen Zhan, and Yiping Cui. 2016. "Enhanced sensitivity of the Z-scan technique on saturable absorbers using radially polarized beams." *Journal of Applied Physics* 119(7): 073103.

Liu, Han, Adam T. Neal, Zhen Zhu, Zhe Luo, Xianfan Xu, David Tománek, and Peide D Ye. 2014. "Phosphorene: an unexplored 2D semiconductor with a high hole mobility." *ACS Nano* 8(4): 4033–4041.

Liu, Lei, Y. P. Feng, and Z. X. Shen. 2003. "Structural and electronic properties of h-BN." *Physical Review B* 68(10): 104102.

Liu, Na, Paul Kim, Ji Heon Kim, Jun Ho Ye, Sunkook Kim, and Cheol Jin Lee. 2014. "Large-area atomically thin MoS2 nanosheets prepared using electrochemical exfoliation." *ACS Nano* 8(7): 6902–6910.

Liu, Sijie, Nengjie Huo, Sheng Gan, Yan Li, Zhongming Wei, Beiju Huang, Jian Liu, Jingbo Li, and Hongda Chen. 2015. "Thickness-dependent Raman spectra, transport properties and infrared photoresponse of few-layer black phosphorus." *Journal of Materials Chemistry C* 3(42): 10974–10980.

Liu, Te-Huan, and Chien-Cheng Chang. 2015. "Anisotropic thermal transport in phosphorene: effects of crystal orientation." *Nanoscale* 7(24): 10648–10654.

Lotya, Mustafa, Yenny Hernandez, Paul J. King, Ronan J. Smith, Valeria Nicolosi, Lisa S. Karlsson, Fiona M. Blighe, Sukanta De, Zhiming Wang, I. T. McGovern, Georg S. Duesberg, and Jonathan N. Coleman. 2009. "Liquid phase production of graphene by exfoliation of graphite in surfactant/water solutions." *Journal of the American Chemical Society* 131(10): 3611–3620.

Low, Tony, A. S. Rodin, A. Carvalho, Yongjin Jiang, Han Wang, Fengnian Xia, and A. H. Castro Neto. 2014. "Tunable optical properties of multilayer black phosphorus thin films." *Physical Review B* 90(7): 075434.

Lu, S. B., L. L. Miao, Z. N. Guo, X. Qi, C. J. Zhao, H. Zhang, S. C. Wen, D. Y. Tang, and D. Y. Fan. 2015. "Broadband nonlinear optical response in multi-layer black phosphorus: an emerging infrared and mid-infrared optical material." *Optics Express* 23(9): 11183–11194.

Luo, Zhe, Jesse Maassen, Yexin Deng, Yuchen Du, Richard P Garrelts, Mark S. Lundstrom, D. Ye Peide, and Xianfan Xu. 2015. "Anisotropic in-plane thermal conductivity observed in few-layer black phosphorus." *Nature Communications* 6: 8572.

Mikhailov, Sergey A. 2014. "Quantum theory of third-harmonic generation in graphene." *Physical Review B* 90(24): 241301.

Nasari, Hadiseh, and Mohammad Sadegh Abrishamian. 2015. "Quasi-phase matching for efficient long-range plasmonic third-harmonic generation via graphene." *Optics Letters* 40(23): 5510–5513.

Nicolosi, Valeria, Manish Chhowalla, Mercouri G. Kanatzidis, Michael S. Strano, and Jonathan N. Coleman. 2013. "Liquid exfoliation of layered materials." *Science* 340(6139): 1226419.

Nikolaidis, Alexandros K., Dimitris S. Achilias, and George P. Karayannidis. 2010. "Synthesis and characterization of PMMA/organomodified montmorillonite nanocomposites prepared by in situ bulk polymerization." *Industrial & Engineering Chemistry Research* 50(2): 571–579.

Nilges, Tom, Marcel Kersting, and Thorben Pfeifer. 2008. "A fast low-pressure transport route to large black phosphorus single crystals." *Journal of Solid State Chemistry* 181(8): 1707–1711.

Novoselov, Kostya S., Andre K. Geim, S. V. Morozov, Da Jiang, Michail I. Katsnelson, I. V. Grigorieva, S. V. Dubonos, and A. A. Firsov. 2005. "Two-dimensional gas of massless Dirac fermions in graphene." *Nature* 438(7065): 197.

O'Neill, Arlene, Umar Khan, Peter N. Nirmalraj, John Boland, and Jonathan N. Coleman. 2011. "Graphene dispersion and exfoliation in low boiling point solvents." *The Journal of Physical Chemistry C* 115(13): 5422–5428.

Ong, Zhun-Yong, Yongqing Cai, Gang Zhang, and Yong-Wei Zhang. 2014. "Strong thermal transport anisotropy and strain modulation in single-layer phosphorene." *The Journal of Physical Chemistry C* 118(43): 25272–25277.

Pálfalvi, L., B. C. Tóth, G. Almási, J. A. Fülöp, and J. Hebling. 2009. "A general Z-scan theory." *Applied Physics B* 97(3): 679.

Parvez, Khaled, Zhong-Shuai Wu, Rongjin Li, Xianjie Liu, Robert Graf, Xinliang Feng, and Klaus Müllen. 2014. "Exfoliation of graphite into graphene in aqueous solutions of inorganic salts." *Journal of the American Chemical Society* 136(16): 6083–6091.

Passaglia, Elisa, Francesca Cicogna, Giulia Lorenzetti, Stefano Legnaioli, Maria Caporali, Manuel Serrano-Ruiz, Andrea Ienco, and Maurizio Peruzzini. 2016. "Novel polystyrene-based nanocomposites by phosphorene dispersion." *RSC Advances* 6(59): 53777–53783.

Passaglia, Elisa, Francesca Cicogna, Federica Costantino, Serena Coiai, Stefano Legnaioli, Giulia Lorenzetti, Silvia Borsacchi, Marco Geppi, Francesca Telesio, Stefan Heun, Andrea Ienco, Manuel Serrano-Ruiz, and Maurizio Peruzzini. 2018. "Polymer-based black phosphorus (bP) hybrid materials by in situ radical polymerization: an effective tool to exfoliate bP and stabilize bP nanoflakes." *Chemistry of Materials* 30(6): 2036–2048.

Pravst, Igor. 2011. "Risking public health by approving some health claims? The case of phosphorus." *Food Policy* 36(5): 726–728.

Qiao, Jingsi, Xianghua Kong, Zhi-Xin Hu, Feng Yang, and Wei Ji. 2014. "High-mobility transport anisotropy and linear dichroism in few-layer black phosphorus." *Nature Communications* 5: 4475.

Qin, Guangzhao, Qing-Bo Yan, Zhenzhen Qin, Sheng-Ying Yue, Hui-Juan Cui, Qing-Rong Zheng, and Gang Su. 2014. "Hinge-like structure induced unusual properties of black phosphorus and new strategies to improve the thermoelectric performance." *Scientific Reports* 4: 6946.

Qin, Guangzhao, Qing-Bo Yan, Zhenzhen Qin, Sheng-Ying Yue, Ming Hu, and Gang Su. 2015. "Anisotropic intrinsic lattice thermal conductivity of phosphorene from first principles." *Physical Chemistry Chemical Physics* 17(7): 4854–4858.

Qiu, M., Z. T. Sun, D. K. Sang, X. G. Han, H. Zhang, and C. M. Niu. 2017. "Current progress in black phosphorus materials and their applications in electrochemical energy storage." *Nanoscale* 9(36): 13384–13403.

Qiu, Meng, Dou Wang, Weiyuan Liang, Liping Liu, Yin Zhang, Xing Chen, David Kipkemoi Sang, Chenyang Xing, Zhongjun Li, and Biqin Dong. 2018. "Novel concept of the smart NIR-light–controlled drug release of black phosphorus nanostructure for cancer therapy." *Proceedings of the National Academy of Sciences* 115(3): 501–506.

Quereda, Jorge, Pablo San-Jose, Vincenzo Parente, Luis Vaquero-Garzon, Aday J. Molina-Mendoza, Nicolás Agraït, Gabino Rubio-Bollinger, Francisco Guinea, Rafael Roldán, and Andres Castellanos-Gomez. 2016. "Strong modulation of optical properties in black phosphorus through strain-engineered rippling." *Nano Letters* 16(5): 2931–2937.

Ren, Xiaohui, Zhongjun Li, Zongyu Huang, David Sang, Hui Qiao, Xiang Qi, Jianqing Li, Jianxin Zhong, and Han Zhang. 2017. "Environmentally robust black phosphorus nanosheets in solution: application for self-powered photodetector." *Advanced Functional Materials* 27(18): 1606834.

Ren, Xiaoyong, Xiao Yang, Guoxin Xie, and Jianbin Luo. 2020. "Black phosphorus quantum dots in aqueous ethylene glycol for macroscale superlubricity." *ACS Applied Nano Materials* 3(5): 4799–4809.

Ribeiro, Henrique B., Marcos A. Pimenta, Christiano J. S. De Matos, Roberto Luiz Moreira, Aleksandr S. Rodin, Juan D. Zapata, Eunézio A. T. De Souza, and Antonio H. Castro Neto. 2015. "Unusual angular dependence of the Raman response in black phosphorus." *ACS Nano* 9(4): 4270–4276.

Rodin, A. S., A. Carvalho, and A. H. Castro Neto. 2014. "Strain-induced gap modification in black phosphorus." *Physical Review Letters* 112(17): 176801.

Rudenko, A. N., and M. I. Katsnelson. 2014. "Quasiparticle band structure and tight-binding model for single- and bilayer black phosphorus." *Physical Review B* 89(20): 201408.

Saito, Yu, and Yoshihiro Iwasa. 2015. "Ambipolar insulator-to-metal transition in black phosphorus by ionic-liquid gating." *ACS Nano* 9(3): 3192–3198.

Schleberger, Marika, and Jani Kotakoski. 2018. "2D material science: defect engineering by particle irradiation." *Materials* 11(10): 1885.

Seol, Jae Hun, Insun Jo, Arden L. Moore, Lucas Lindsay, Zachary H. Aitken, Michael T. Pettes, Xuesong Li, Zhen Yao, Rui Huang, and David Broido. 2010. "Two-dimensional phonon transport in supported graphene." *Science* 328(5975): 213–216.

Shao, Jundong, Hanhan Xie, Hao Huang, Zhibin Li, Zhengbo Sun, Yanhua Xu, Quanlan Xiao, Xue-Feng Yu, Yuetao Zhao, and Han Zhang. 2016. "Biodegradable black phosphorus-based nanospheres for in vivo photothermal cancer therapy." *Nature Communications* 7: 12967.

Sheik-Bahae, Mansoor, Ali A. Said, and Eric W. Van Stryland. 1989. "High-sensitivity, single-beam n 2 measurements." *Optics Letters* 14(17): 955–957.

Sheik-Bahae, Mansoor, Ali A. Said, T.-H. Wei, David J. Hagan, and Eric W. Van Stryland. 1990. "Sensitive measurement of optical nonlinearities using a single beam." *IEEE Journal of Quantum Electronics* 26(4): 760–769.

Slack, Glen A. 1965. "Thermal conductivity of elements with complex lattices: B, P, S." *Physical Review* 139(2A): A507.

Smallwood, Ian. 2012. *Handbook of organic solvent properties.* Oxford: Butterworth-Heinemann.

Smith, Joshua B., Daniel Hagaman, and Hai-Feng Ji. 2016. "Growth of 2D black phosphorus film from chemical vapor deposition." *Nanotechnology* 27(21): 215602.

Song, Yufeng, Si Chen, Qian Zhang, Lei Li, Luming Zhao, Han Zhang, and Dingyuan Tang. 2016. "Vector soliton fiber laser passively mode locked by few layer black phosphorus-based optical saturable absorber." *Optics Express* 24(23): 25933–25942.

Sun, Jie, Hyun-Wook Lee, Mauro Pasta, Hongtao Yuan, Guangyuan Zheng, Yongming Sun, Yuzhang Li, and Yi Cui. 2015. "A phosphorene–graphene hybrid material as a high-capacity anode for sodium-ion batteries." *Nature Nanotechnology* 10(11): 980.

Sun, Zhengbo, Hanhan Xie, Siying Tang, Xue-Feng Yu, Zhinan Guo, Jundong Shao, Han Zhang, Hao Huang, Huaiyu Wang, and Paul K. Chu. 2015a. "Ultrasmall black phosphorus quantum dots: synthesis and use as photothermal agents." *Angewandte Chemie International Edition* 54(39): 11526–11530.

Sun, Zhengbo, Hanhan Xie, Siying Tang, Xue-Feng Yu, Zhinan Guo, Jundong Shao, Han Zhang, Hao Huang, Huaiyu Wang, and Paul K. Chu. 2015b. "Inside back cover: ultrasmall black phosphorus quantum dots: synthesis and use as photothermal agents." *Angewandte Chemie International Edition* 54(39): 11581–11581.

Sun, Zhengbo, Yuetao Zhao, Zhibin Li, Haodong Cui, Yayan Zhou, Weihao Li, Wei Tao, Han Zhang, Huaiyu Wang, Paul K. Chu, and Xue-Feng Yu. 2017. "TiL4-coordinated black phosphorus quantum dots as an efficient contrast agent for in vivo photoacoustic imaging of cancer." *Small* 13(11): 1602896.

Suryawanshi, Sachin R., Satya N. Guin, Arindom Chatterjee, Vikas Kashid, Mahendra A. More, Dattatray J. Late, and Kanishka Biswas. 2016. "Low frequency noise and photo-enhanced field emission from ultrathin PbBi2Se4 nanosheets." *Journal of Materials Chemistry C* 4(5): 1096–1103.

Suryawanshi, Sachin R., Mahendra A. More, and Dattatray J. Late. 2016. "Laser exfoliation of 2D black phosphorus nanosheets and their application as a field emitter." *RSC Advances* 6(113): 112103–112108.

Tian, Bin, Bining Tian, Bethany Smith, M. C. Scott, Qin Lei, Ruinian Hua, Yue Tian, and Yi Liu. 2018. "Facile bottom-up synthesis of partially oxidized black phosphorus nanosheets as metal-free photocatalyst for hydrogen evolution." *Proceedings of the National Academy of Sciences* 115(17): 4345.

Tran, Vy, Ryan Soklaski, Yufeng Liang, and Li Yang. 2014. "Layer-controlled band gap and anisotropic excitons in few-layer black phosphorus." *Physical Review B* 89(23): 235319.

Vogt, Patrick, Paola De Padova, Claudio Quaresima, Jose Avila, Emmanouil Frantzeskakis, Maria Carmen Asensio, Andrea Resta, Bénédicte Ealet, and Guy Le Lay. 2012. "Silicene: compelling experimental evidence for graphenelike two-dimensional silicon." *Physical Review Letters* 108(15): 155501.

Walia, Sumeet, Ylias Sabri, Taimur Ahmed, M. Field, Rajesh Ramanathan, Aram Arash, S. Bhargava, Sharath Sriram, Madhu Bhaskaran, and Vipul Bansal. 2016. "Defining the role of humidity in the ambient degradation of few-layer black phosphorus." *2D Materials* 4(1): 1–8.

Walia, Sumeet, Sivacarendran Balendhran, Taimur Ahmed, Mandeep Singh, Christopher El-Badawi, Mathew D Brennan, Pabudi Weerathunge, Md Nurul Karim, Fahmida Rahman, and Andrea Rassell. 2017. "Ambient protection of few-layer black phosphorus via sequestration of reactive oxygen species." *Advanced Materials* 29(27): 1700152.

Wang, Hui, Xianzhu Yang, Wei Shao, Shichuan Chen, Junfeng Xie, Xiaodong Zhang, Jun Wang, and Yi Xie. 2015. "Ultrathin black phosphorus nanosheets for efficient singlet oxygen generation." *Journal of the American Chemical Society* 137(35): 11376–11382.

Wang, Kangpeng, Beata M. Szydłowska, Gaozhong Wang, Xiaoyan Zhang, Jing Jing Wang, John J. Magan, Long Zhang, Jonathan N Coleman, Jun Wang, and Werner J. Blau. 2016. "Ultrafast nonlinear excitation dynamics of black phosphorus nanosheets from visible to mid-infrared." *ACS Nano* 10(7): 6923–6932.

Wang, Xiaomu, and Shoufeng Lan. 2016. "Optical properties of black phosphorus." *Advances in Optics and Photonics* 8(4): 618–655.

Wang, X., A. M. Jones, K. L. Seyler, V. Tran, Y. Jia, H. Zhao, H. Wang, L. Yang, X. Xu, and F. Xia. 2015. "Highly anisotropic and robust excitons in monolayer black phosphorus." *Nature Nanotechnology* 10(6): 517–521.

Wang, Yingwei, Guanghui Huang, Haoran Mu, Shenghuang Lin, Jiazhang Chen, Si Xiao, Qiaoliang Bao, and Jun He. 2015. "Ultrafast recovery time and broadband saturable absorption properties of black phosphorus suspension." *Applied Physics Letters* 107(9): 091905.

Wang, Yuxi, Ning Xu, Deyu Li, and Jia Zhu. 2017. "Thermal properties of two dimensional layered materials." *Advanced Functional Materials* 27(19): 1604134.

Wei, Ning, Cunjing Lv, and Zhiping Xu. 2014. "Wetting of graphene oxide: a molecular dynamics study." *Langmuir* 30(12): 3572–3578.

Wei, Qun, and Xihong Peng. 2014. "Superior mechanical flexibility of phosphorene and few-layer black phosphorus." *Applied Physics Letters* 104(25): 251915.

Wood, Joshua D., Spencer A. Wells, Deep Jariwala, Kan-Sheng Chen, EunKyung Cho, Vinod K. Sangwan, Xiaolong Liu, Lincoln J. Lauhon, Tobin J. Marks, and Mark C. Hersam. 2014. "Effective passivation of exfoliated black phosphorus transistors against ambient degradation." *Nano Letters* 14(12): 6964–6970.

Woodward, R. I., R. T. Murray, C. F. Phelan, R. E. P. De Oliveira, T. H. Runcorn, E. J. R. Kelleher, S. Li, E. C. De Oliveira, G. J. M. Fechine, and G. Eda. 2016. "Characterization of the second-and third-order nonlinear optical susceptibilities of monolayer MoS2 using multiphoton microscopy." *2D Materials* 4(1): 011006.

Woomer, Adam H., Tyler W. Farnsworth, Jun Hu, Rebekah A. Wells, Carrie L. Donley, and Scott C. Warren. 2015. "Phosphorene: synthesis, scale-up, and quantitative optical spectroscopy." *ACS Nano* 9(9): 8869–8884.

Wu, Jishan, Wojciech Pisula, and Klaus Müllen. 2007. "Graphenes as potential material for electronics." *Chemical Reviews* 107(3): 718–747.

Wu, Shuxing, Kwan San Hui, and Kwun Nam Hui. 2018. "2D black phosphorus: from preparation to applications for electrochemical energy storage." *Advanced Science* 5(5): 1700491.

Wu, Tong, Jinchen Fan, Qiaoxia Li, Penghui Shi, Qunjie Xu, and Yulin Min. 2018. "Palladium nanoparticles anchored on anatase titanium dioxide-black phosphorus hybrids with heterointerfaces: highly electroactive and durable catalysts for ethanol electrooxidation." *Advanced Energy Materials* 8(1): 1701799.

Wu, Y., B. C. Yao, Q. Y. Feng, X. L. Cao, X. Y. Zhou, Y. J. Rao, Y. Gong, W. L. Zhang, Z. G. Wang, and Y. F. Chen. 2015. "Generation of cascaded four-wave-mixing with graphene-coated microfiber." *Photonics Research* 3(2): A64–A68.

Xia, Fengnian, Han Wang, and Yichen Jia. 2014. "Rediscovering black phosphorus as an anisotropic layered material for optoelectronics and electronics." *Nature Communications* 5: 4458.

Xia, Fengnian, Han Wang, Di Xiao, Madan Dubey, and Ashwin Ramasubramaniam. 2014. "Two-dimensional material nanophotonics." *Nature Photonics* 8(12): 899.

Xin, Xiufeng, Fang Liu, Xiao-Qing Yan, Wangwei Hui, Xin Zhao, Xiaoguang Gao, Zhi-Bo Liu, and Jian-Guo Tian. 2018. "Two-photon absorption and non-resonant electronic nonlinearities of layered semiconductor TlGaS 2." *Optics Express* 26(26): 33895–33905.

Xu, Hao, Xiaoyu Han, Zhuangnan Li, Wei Liu, Xiao Li, Jiang Wu, Zhengxiao Guo, and Huiyun Liu. 2018. "Epitaxial growth of few-layer black phosphorene quantum dots on si substrates." *Advanced Materials Interfaces* 5(21): 1801048.

Xu, Yanhua, Zhiteng Wang, Zhinan Guo, Hao Huang, Quanlan Xiao, Han Zhang, and Xue-Feng Yu. 2016. "Solvothermal synthesis and ultrafast photonics of black phosphorus quantum dots." *Advanced Optical Materials* 4(8): 1223–1229.

Yang, Bowen, Junhui Yin, Yu Chen, Shanshan Pan, Heliang Yao, Youshui Gao, and Jianlin Shi. 2018. "2D-black-phosphorus-reinforced 3D-printed scaffolds: a stepwise countermeasure for osteosarcoma." *Advanced Materials* 30(10): 1705611.

Yang, Yong, Shigemasa Matsubara, Liangming Xiong, Tomokatsu Hayakawa, and Masayuki Nogami. 2007. "Solvothermal synthesis of multiple shapes of silver nanoparticles and their SERS properties." *The Journal of Physical Chemistry C* 111(26): 9095–9104.

Yang, Zhibin, Jianhua Hao, Shuoguo Yuan, Shenghuang Lin, Hei Man Yau, Jiyan Dai, and Shu Ping Lau. 2015. "Field-effect transistors based on amorphous black phosphorus ultrathin films by pulsed laser deposition." *Advanced Materials* 27(25): 3748–3754.

Yasaei, Poya, Bijandra Kumar, Tara Foroozan, Canhui Wang, Mohammad Asadi, David Tuschel, J. Ernesto Indacochea, Robert F. Klie, and Amin Salehi-Khojin. 2015. "High-quality black phosphorus atomic layers by liquid-phase exfoliation." *Advanced Materials* 27(11): 1887–1892.

Yeong, K. S., K. H. Maung, and J. T. L. Thong. 2007. "The effects of gas exposure and UV illumination on field emission from individual ZnO nanowires." *Nanotechnology* 18(18): 185608.

You, Xueqiu, Na Liu, Cheol Jin Lee, and James Jungho Pak. 2014. "An electrochemical route to MoS2 nanosheets for device applications." *Materials Letters* 121: 31–35.

Yuan, Hongtao, Xiaoge Liu, Farzaneh Afshinmanesh, Wei Li, Gang Xu, Jie Sun, Biao Lian, Alberto G. Curto, Guojun Ye, and Yasuyuki Hikita. 2015. "Polarization-sensitive broadband photodetector using a black phosphorus vertical p–n junction." *Nature Nanotechnology* 10(8): 707.

Zhang, Jingdi, Xuefeng Yu, Weijia Han, Bosai Lv, Xiaohong Li, Si Xiao, Yongli Gao, and Jun He. 2016. "Broadband spatial self-phase modulation of black phosphorous." *Optics Letters* 41(8): 1704–1707.

Zhang, Saifeng, Ningning Dong, Niall McEvoy, Maria O'Brien, Sinéad Winters, Nina C. Berner, Chanyoung Yim, Yuanxin Li, Xiaoyan Zhang, and Zhanghai Chen. 2015. "Direct observation of degenerate two-photon absorption and its saturation in WS2 and MoS2 monolayer and few-layer films." *ACS Nano* 9(7): 7142–7150.

Zhang, Shuang, Jiong Yang, Renjing Xu, Fan Wang, Weifeng Li, Muhammad Ghufran, Yong-Wei Zhang, Zongfu Yu, Gang Zhang, and Qinghua Qin. 2014. "Extraordinary photoluminescence and strong temperature/angle-dependent Raman responses in few-layer phosphorene." *ACS Nano* 8(9): 9590–9596.

Zhang, Wentao, Yanru Wang, Daohong Zhang, Shaoxuan Yu, Wenxin Zhu, Jing Wang, Fangqing Zheng, Shuaixing Wang, and Jianlong Wang. 2015. "A one-step approach to the large-scale synthesis of functionalized MoS2 nanosheets by ionic liquid assisted grinding." *Nanoscale* 7(22): 10210–10217.

Zhang, Xiao, Haiming Xie, Zhengdong Liu, Chaoliang Tan, Zhimin Luo, Hai Li, Jiadan Lin, Liqun Sun, Wei Chen, Zhichuan Xu, Linghai Xie, Wei Huang, and Hua Zhang. 2015. "Black phosphorus quantum dots." *Angewandte Chemie* 127(12): 3724–3728.

Zhang, Yang, Guanyin Gao, Helen L. W. Chan, Jiyan Dai, Yu Wang, and Jianhua Hao. 2012a. "Piezo-Phototronic Effect-Induced Dual-Mode Light and Ultrasound Emissions from ZnS:Mn/PMN–PT Thin-Film Structures." *Advanced Materials* 24(13): 1729–1735.

Zhang, Yang, Guanyin Gao, Helen L. W. Chan, Jiyan Dai, Yu Wang, and Jianhua Hao. 2012b. "Piezo-phototronic effect-induced dual-mode light and ultrasound emissions from ZnS: Mn/PMN–PT thin-film structures." *Advanced Materials* 24(13): 1729–1735.

Zhang, Ying-Yan, Qing-Xiang Pei, Jin-Wu Jiang, Ning Wei, and Yong-Wei Zhang. 2016. "Thermal conductivities of single-and multi-layer phosphorene: a molecular dynamics study." *Nanoscale* 8(1): 483–491.

Zhang, Yuanbo, Yan-Wen Tan, Horst L. Stormer, and Philip Kim. 2005. "Experimental observation of the quantum Hall effect and Berry's phase in graphene." *Nature* 438(7065): 201.

Zhao, Wancheng, Zhimin Xue, Jinfang Wang, Jingyun Jiang, Xinhui Zhao, and Tiancheng Mu. 2015. "Large-scale, highly efficient, and green liquid-exfoliation of black phosphorus in ionic liquids." *ACS Applied Materials & Interfaces* 7(50): 27608–27612.

Zhao, Yuetao, Huaiyu Wang, Hao Huang, Quanlan Xiao, Yanhua Xu, Zhinan Guo, Hanhan Xie, Jundong Shao, Zhengbo Sun, and Weijia Han. 2016. "Surface coordination of black phosphorus for robust air and water stability." *Angewandte Chemie International Edition* 55(16): 5003–5007.

Zheng, Xin, Runze Chen, Gang Shi, Jianwei Zhang, Zhongjie Xu, and Tian Jiang. 2015. "Characterization of nonlinear properties of black phosphorus nanoplatelets with femtosecond pulsed Z-scan measurements." *Optics Letters* 40(15): 3480–3483.

Zhou, Feng, and Wei Ji. 2017. "Two-photon absorption and subband photodetection in monolayer MoS 2." *Optics Letters* 42(16): 3113–3116.

Zhou, Qionghua, Qian Chen, Yilong Tong, and Jinlan Wang. 2016. "Light-induced ambient degradation of few-layer black phosphorus: mechanism and protection." *Angewandte Chemie International Edition* 55(38): 11437–11441.

Zhu, Jian, Joohoon Kang, Junmo Kang, Deep Jariwala, Joshua D. Wood, Jung-Woo T. Seo, Kan-Sheng Chen, Tobin J. Marks, and Mark C. Hersam. 2015. "Solution-processed dielectrics based on thickness-sorted two-dimensional hexagonal boron nitride nanosheets." *Nano Letters* 15(10): 7029–7036.

Zhu, Jie, Haechan Park, Jun-Yang Chen, Xiaokun Gu, Hu Zhang, Sreejith Karthikeyan, Nathaniel Wendel, Stephen A Campbell, Matthew Dawber, and Xu Du. 2016. "Revealing the origins of 3D anisotropic thermal conductivities of black phosphorus." *Advanced Electronic Materials* 2(5): 1600040.

Zhu, Zhen, and David Tománek. 2014. "Semiconducting layered blue phosphorus: a computational study." *Physical Review Letters* 112(17): 176802.

Ziletti, A., A. Carvalho, D. K. Campbell, D. F. Coker, and A. H. Castro Neto. 2015. "Oxygen defects in phosphorene." *Physical Review Letters* 114(4): 046801.

4 Composites and Heterostructures of Black Phosphorus

4.1 INTRODUCTION

Layer-by-layer-assembled 2D nanosheet heterostructures are an important class of materials because they have novel hybrid physicochemical properties and applications. Heterostructures are modern hybrid materials that combine the special intrinsic properties of various 2D materials (Geim and Grigorieva 2013; Pomerantseva and Gogotsi 2017). In general, heterostructures are produced by layer-by-layer mechanical transfer (Chen et al. 2008; Choi et al. 2013) chemical vapor deposition (CVD) (Levendorf et al. 2012; Liu et al. 2013), solvothermal processes (Shang et al. 2009; Wang et al. 2020), and wet chemical reactions to assemble different nanomaterials through covalent bond formation or noncovalent interactions. Different nanomaterials are attached to each other in covalently formed heterostructures through direct bond formation or organic linker bonding. Different nanomaterials, on the other hand, are stacked layer by layer to form noncovalent heterostructures through van der Waals interactions or electrostatic interactions. Black phosphorous (BP) logically and possibly practically has reached a mature stage after many years of intensive effort. Among the 2D materials, BP is considered one of the important components in many vdWs heterostructures assembly due to its high carrier mobility, strong environmental stability, tunable work function, and mature processing technique. Moreover, the possibility of combining BP with another type of 2D layered crystals such as graphene and 2D transition metal dichalcogenides (TMDCs) and polymers as a result of van der Waals has expanded this field dramatically. As a result of advancement in the area of heterostructures (a combination of BP with other 2D material), more and more research groups increasingly produced sophisticated devices and materials. On increasing the number of 2D crystals, parameters, and sequences for consideration, the choice of possible van der Waals structures is limited and it might cause a snowball effect. Additionally, it might be a time-consuming process to explore the huge parametric study and relate this work with the 2D layered material such as graphene and TMDC-based structures that have been shown to be stable. Recently research on devices and semiconductor heterostructures may serve as a guide to judge the probable durability of research on van der Waals materials. In this chapter, the detail of composites of BP and its heterostructures with other 2D layered material are discussed.

4.2 COMPOSITES OF BLACK PHOSPHORUS WITH POLYMERS

During the past five years, BP-based structures have been widely used in the field of electronic, photonic, and biomedical applications. Although there have been a lot of success stories, the instability of BP as a result of interaction in the ambient atmosphere (with moisture or oxygen-containing reactions) has limited its practical industrial applications. Xu et al. (2017) used an electrospinning-based technique and developed a facile methodology to significantly stabilize the black phosphorus quantum dots (BPQDs), as a result producing an unchanging and smooth polymethyl methacrylate (PMMA) functionalized BPQDs-based composite nanofiber film. To develop such a smooth film, the pure form of polymeric (PMMA) combination with (BPQDs) solution was homogenized at a temperature of 60°C under vigorous mechanical stirring for six hours to make a homogenous solution. After that in a typical well-equipped electrospinning machine, the homogeneous solution of PMMA and BPQDs was magnificently electro-spun into a smooth and well-organized film of composite nanofiber. The PMMA functionalized BPQDs-based composite nanofiber film had the broadband nonlinear optical saturable absorption spectrum (a range from the visible band to the mid-IR band). Mu et al. (2015) anticipate two different methodologies to produce BP-polymer composite material that can minimize the transferring problem of composite film and easily be transported onto the facets of optical fiber working as a photonic device.

In the first methodology study, a sandwich-like structure of two pieces of PMMA in which BP was placed between thin films was structured, as in Figure 4.1a. The mechanical exfoliation technique was used to peel the resulting PMMA-BP-PMMA composite film for further optical studies. Although this technique is simple, the reproducibility and uniform dispersion of BP material on the PMMA surface is a big problem. To keep in mind the above-mentioned problem, the author has used facile and robust networked membranes involving electro-spun BP/PVP nanofibers (Figure 4.1b). Figure 4.1c shows the uniformly aligned surface morphology characterization of electro-spun fiber-based thin film characterized by scanning electron microscope (SEM) and the measurement of the fiber's uniform diameter around 200 nm as shown in the high magnification SEM image in Figure 4.1d.

Transmission electron microscopy (TEM) is used to analyze composites' internal morphology and the dark dissimilarity in the nanofiber (TEM pictures in Figure 4.1e), providing the pieces of evidence for the effective encapsulation process of BP in PVP. Moreover, Figure 4.1f shows the absorption spectra of a reference PVP film and BP-PVP composite and transparent uniform composite thin film produced as a result of the electrospinning process. BP displays a rapid carrier dynamic and the composite of polymer functionalized BP has a variation depth value of ~10.6 percent. Moreover, a highly stable Q-switched value for pulse generation as well as the single pulse energy of 194 nJ was accomplished. As a result of BP polymer composite, the polymer protects BP and keeps it away from degradation in ambient environments. Additionally, the functionalization of polymer on BP surface also produces a novel optical-based material with the exceptional optical characteristics of BP and the structural-based properties of the polymeric material (Mu et al. 2015). Li et al. (2016) introduced the liquid phase exfoliation-based approach (LPE) and a solvent interchange procedure and recommended a cost-effective, simple and industrially appropriate method for the

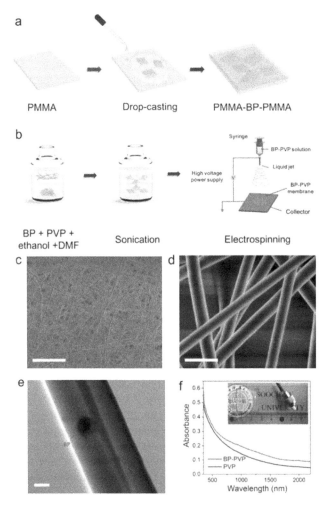

FIGURE 4.1 a) The fabrication process of a sandwich-like structure of two pieces of PMMA in which BP was placed between thin films. b) Flexible and bendable conductive thin film is consequently designed, which can readily be peeled off from the surface of the PET sheet. c&d) Surface morphology monitoring (SEM pictures) of PVP functionalized BP nanocomposite-based membrane made up as a result of the electrospinning process. The length of scale bars of SEM images is marked as c) 5 μm and d) 500 nm, respectively. e) The high magnification TEM picture of PVP functionalized BP nanofiber composite shows the internal structure and morphology of nanofibers. The scale bar on an image is 50 nm. The arrow designates the presence of BP nanoflake on the surface of polymer fibers. f) The UV visible absorption spectra of a reference PVP film (lower line) and PVP functionalized BP composite (grey line) and transparent uniform composite thin film produced as a result of the electrospinning process. The photograph of PVP functionalized BP composite-based membrane prepared as a result of electrospinning is shown on (inset) a quartz-based substrate (20×20 mm, left) and the optical fiber ferrule (right). Adapted with permission from Mu et al. (2015), Copyright © 2015, WILEY–VCH Verlag GmbH & Co. KGaA, Weinheim.

preparation of BP-based nanomaterial. After the passivation with the polycarbonate, the process chemical inertness of FL-BP under ambient reaction atmosphere/conditions was gradually improved and used as a stable material (environmentally stable) with the optical-based transparency moved from visible to the infrared range.

Additionally, large energy value for pulse creation at 1.55 μm telecommunication wavelength, the PC functionalized FL-BP-based composite film was also consequently used as an SA. A Q-switched laser with a pulse duration down to 1.65 seconds and pulse energy up to 25.2 nJ is attained at a low pump power of 71.7 mW. These findings can be boosted for scalable photonic applications and further research in electronics and photonics, where the FL-BP based environmentally stable devices are required.

Luo et al. (2018) introduced a one-step facile and robust electrochemical based deposition technique and established a conducting Polypyrrole (PPy) functionalized BP self-standing composite film for enhancing the capacitance as well as the value of cycling stability. For the preparation of the composite film, at the initial stage using the sonication process with vigorous stirring for the dispersion of PPy monomer and nanosheets of BP into the electrolyte solution, this made it easy to capture the BP nanosheets by PPy chains during the process of electrochemical polymerization (EP) as shown in Figure 4.2a.

Afterward, the constant-voltage deposition technique was used for this PPy enveloped (OH-protected) BP-based hybrids dispersed/coating onto the surface of the Indium tin oxide (ITO) glass substrate. After a continued electrodeposition process, a flexible and bendable conductive thin film is consequently designed, which can readily be peeled off from the surface of the PET sheet, as shown in Figure 4.2b. The high value of capacitance 452.8 F/g (7.7 F/cm^3), a suitable value of mechanical based flexibility after 3000 bending and outstanding value of cycling stability after 10,000 charges/discharge cycles are presented by the solid-state flexible supercapacitor (SC) assembled as a result of PPy/BP based composite uniform film.

It is believed that the large surface area of self-standing PPy functionalized BP composite film (Figure 4.2c) is highly desired for high-performance flexible SC and might be an encouraging option for the energy storage device requirements of a diversity of portable electronics, motivating additional ingenious arrangements of 2D materials and conductive polymer with improved cyclic life and capacitance (Luo et al. 2018).

4.3 HETEROSTRUCTURES OF BLACK PHOSPHORUS WITH GRAPHENE

Recently BP has attracted a lot of interest from researchers due to its promising application as an anode material for sodium-ion batteries (SIBs), due to its good electrical property (300 S/m) and high value of theoretical capacity (2596 mAh/g). In contrast, a large variation in volume during the electrochemical cycling process makes it problematic for practical applications (Liu et al. 2017; Zhang et al. 2013; Li, Wang, et al. 2014). So it's time to develop some heterostructures of BP with some other 2D layered material such as graphene and transition metal dichalcogenides

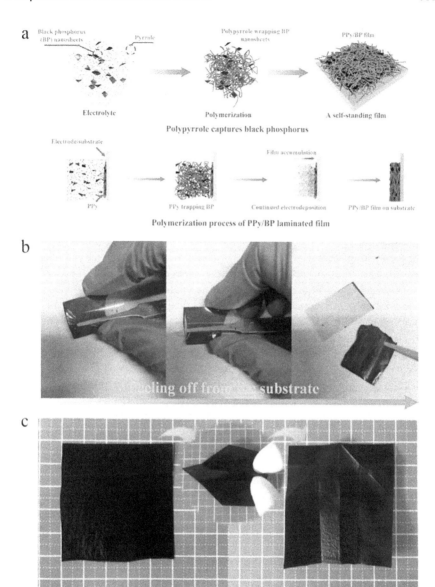

FIGURE 4.2 a) The manufacturing of the composite film, at the initial stage using the sonication process with vigorous stirring for the dispersion the monomer of PPy and nanosheets of BP into the electrolyte solution, made it easy to capture the BP nanosheets by PPy chains during the process of electrochemical polymerization. b) A flexible conductive thin PPy functionalized BP film after a continued electrodeposition procedure was consequently designed, which can readily be peeled off from the PET sheet. c) Snapshot of the huge area self-standing PPy functionalized BP composite film (having square size 5 cm × 5 cm) is highly desired for high-performance flexible super capacitors. Adapted with permission from Luo et al. (2018), Copyright © 2018, American Chemical Society.

(TMDCs) to overcome the problem. Liu et al. (2017) introduced the BP based electrode and developed a normal preparation of 4-nitrobenzene-diazonium (4-NBD) having a chemical attachment with the reduced graphene oxide (RGO) based hybrid (4-RBP), improved the performance of sodium-ion battery (SIB). The resultant product is responsible for raising the value for the thickness of BP nanosheet as a result of the coating of RGO layers on the BP surface due to P-C bond and P-O-C bond (Figure 4.3a–b). The network between RGO and BP develops greater as a result of the alteration of 4-NBD and the resultant storing ability of sodium ions increases. Due to the functionalization of BP on graphene, the alterable performance efficiency of BP-based electrodes in SIB is gradually increased. As a result of the strong interaction between the graphene and chemical functionalized BP, there is an improvement in the stability process of BP during the working of SIB. Meanwhile, the thickness of the bridging increases for BP. The surface energy existing for broadening the network between graphene and BP nanosheet reduces, and as a result the cycling storage performance of sodium ions due to enlarged channels is improved. As a result of the bridging, the thickness of BP increases, and the surface energy available reduces (Figures 4.3c–g) (Liu et al. 2017).

The BP-based SIB anode displays a specific capacity value of 1472 mAh/ g at 0.1A g^{-1} in the 50th cycle and it is gradually decreased to 650 mAh g^{-1} at 1 Ag^{-1} after enhancing the number of cycles to 200. In another study for the combination of graphene and phosphorus (PG), Zhang et al. (2013) proposed a low-cost, facile, and scalable thermal annealing method using triphenylphosphine (TPP) and graphite oxide (GO) as carbon and phosphorus sources, respectively. The resultant PG showed outstanding tolerance to methanol crossover effect, remarkable catalytic activity, and outstanding stability for a long time and represented as an efficient metal-free electrocatalyst in oxidation-reduction reaction (ORR). Moreover, by adding the small content of carbon black (CB) to transform its conductivity, the oxidation-reduction reaction ORR activity of PG could be further improved and employed as an anode material in lithium-ion batteries (LIBs).

As a result of doping with carbon, electrochemical characteristics are significantly improved compared to undoped graphene. LIBs and ORR findings have both recommended the enormous potential of PG and display highly improved cycle and rate competencies relative to undoped material of graphene. The experimental results provide an innovative graphene-based material that transports useful evidence to investigate further and prepare phosphorus-doped carbon materials, as well as for advanced energy conversion and storage applications such as functional for both LIBs and ORR (Zhang et al. 2013).

4.4 HETEROSTRUCTURES OF BLACK PHOSPHORUS WITH TRANSITION METAL DICHALCOGENIDES

A single layer of TMDCs—such as tungsten disulfide (WS_2) molybdenum disulfide ($MoSe_2$), and tungsten diselenide (WSe_2)—has the value of direct bandgaps more than 1 eV. Additionally, the non-zero bandgap make the monolayer characteristics of phosphorene and transition metal disulfides exceed the graphene too much in optoelectronic devices and electronic-based applications (Tian et al. 2016).

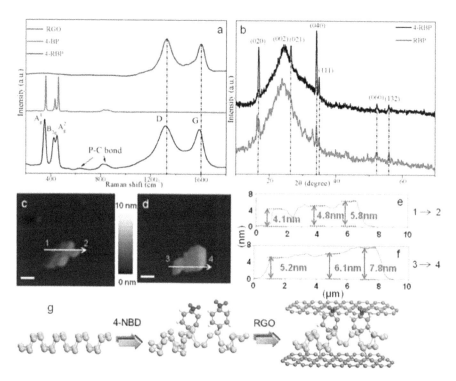

FIGURE 4.3 a&b) The Raman spectra and XRD patterns of BP resultant product responsible for growing the width value of BP nanosheet due to the coating of RGO layers of spread on the surface of BP by P-C bond and P-O-C bond. c–f) AFM pictures of BP nanoflake and 4-BP shows the thickness and surface morphology. g) Molecular model of bonding with reduced graphene oxide (RGO) and 4-NBD alteration. Adapted with permission from Liu et al. (2017), Copyright © 2017, American Chemical Society.

4.4.1 HETEROSTRUCTURES OF BLACK PHOSPHORUS WITH MOS$_2$

To achieve the desired and precise electronic or optoelectronic properties, the isolation of different 2D layered materials such as BP, TMDCs, and graphene has raised the opportunity of designing and developing novel van der Waals (vdWs)-based heterostructures (Jiang et al. 2015). Huang et al. (2015) equate single-layer BP band structures to confirm the interaction of vdWs interlayer forces for monolayer MoS2 and bilayer BP/MoS$_2$. The band structures value of monolayers, the predictable band structure, and band alignments of hybrid BP and MoS$_2$ are shown in Figure 4.4a–b.

The resultant electronic-based structures of both the MoS$_2$ and the BP layers are well-conserved and after they combine in a bilayer form of the heterostructure, the bandgaps of monolayer MoS$_2$ and BP decreased. Deng et al. (2014) explain that the small bandgap of BP and MoS$_2$ heterostructures significantly impact the electrical properties of resultant heterostructures. Moreover, they analyzed a gate-tunable single-layer MoS$_2$ and BP van der Waals heterojunction pin diode, and these diodes showed a strong current-resolving I-V feature (Figure 4.5a–b). It is very appropriate

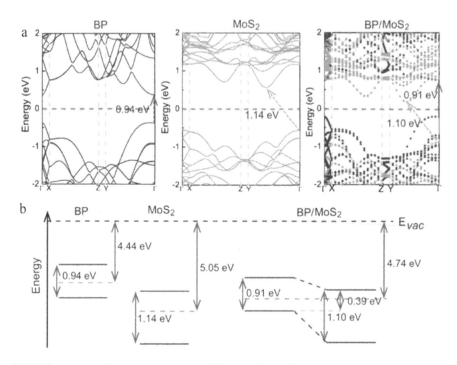

FIGURE 4.4 a) The band structures of SL-BP, SL-MoS$_2$, and hybrid bilayer BP/MoS$_2$ structure. The Fermi level is adjusted as zero. b) Band alignments of SL-BP, SL-MoS$_2$, and bilayer hybrid BP/MoS$_2$. The dotted lines are considered as the Fermi level and the vacuum level is taken as reference. Adopted with permission from Huang et al. (2015), Copyright © 2015, American Chemical Society.

for broadband and sensitive photodetection processes, possibly in narrow bandgap values and high carrier mobility of FL-BP. Gate-tunable pn diode based on an n-type monolayer MoS$_2$/p-type BP concludes a van der Waals pn heterojunction with a dark purple area is single-layer MoS$_2$, while the blue flake is FL-BP (Figure 4.5c). These ultrathin pn diodes at the wavelength of 633 nm show a maximum photodetection of 418 mA/W upon illumination and photovoltaic energy conversion with an external quantum efficiency of 0.3 percent.

These pn diodes show promise for solar energy harvesting and broadband photodetection. Chen et al. (2015) fabricated MoS$_2$–BP heterojunction based devices in another study. Due to Fermi level (unpinned) of BP and the narrow bandgap and by the electrostatic gating this heterojunction process might be tuned to either n–n or p–n. Moreover, in both of the junctions (p–n and n–n), the transformation of alternating current to direct current behaviors were observed. The energy barrier is shaped at the boundary of narrow bandgap BP and wide bandgap MoS$_2$ is responsible for the rectification process of the MoS$_2$–BP n–n junction was recognized. At different thickness scale, the gating requirement of reverse current, forwarding current, and the conversion of alternating current to direct current behaviors of the heterojunction were

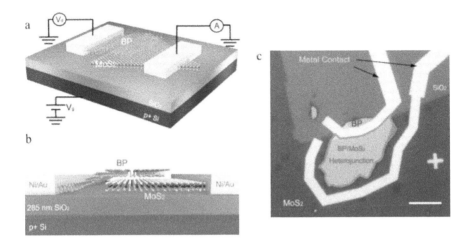

FIGURE 4.5 a&b) Gate-tunable single-layer BP and MoS$_2$ van der Waals structures pin diode and these diodes showed strong current-resolving I-V features. Voltage (Vd) is useful across the device during the electrical measurements. The bias voltage (Vg) is functional in the back gate direction. c) In the optical microscopic image of the prepared device, gate-tunable pn diode based on an n-type single-layer MoS$_2$ / p-type BP concludes a van der Waals pn heterojunction with single-layer MoS$_2$. In contrast, the color like blue flake is FL-BP. Scale bar, 10 µm. Adapted with permission from Deng et al. (2014), Copyright © 2014, American Chemical Society.

systematically studied and it was suggested that after proper scheming of the layer thickness value of BP and MoS$_2$ flake, the electrical based stuff of the heterojunction might be organized (Chen et al. 2015).

4.4.2 Heterostructures of Black Phosphorus with WS$_2$

Hydrogen (H$_2$) is a green fuel (zero carbon emission fuel), which means it is a clean and renewable energy source. To control the high price of fuel, effective photocatalysts that use the earth's abundant elements as their basic source must be developed (Zhu et al. 2018; Zhang, Lan, and Wang 2016; Wang et al. 2009; Hu, Shen, and Jimmy 2017; Ansari et al. 2016). The photo-generated electrons from excited BP are trapped by 2D tungsten disulfide (WS$_2$) and as a result play a critical role in the development of the performance of the photocatalyst, causing water reduction to H$_2$ (Zhu et al. 2018).

Zhu et al. (2018) studied the combination of WS$_2$ and 2D hybrid material of BP using an NMP based solvent exfoliation approach and applied as a stable photocatalyst (in the absence of noble metal) under NIR light radiance for H$_2$ production for the first time. TEM and Atomic force microscopy (AFM) are applied to study the fine morphologies of as-prepared WS$_2$ as well as nanoflakes of BP. TEM images (Figure 4.6a–b) show the internal morphology of 2D nanoflake-like structures with controllable size (WS$_2$ size from 50 to 100 nm and BP with a size 200 nm^{-1} µm). The high-resolution

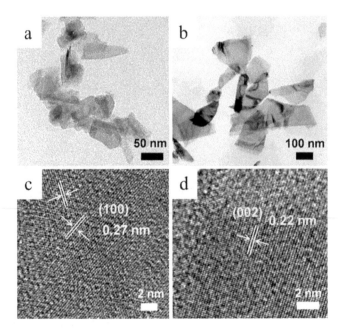

FIGURE 4.6 The internal morphology of WS$_2$ and BP nanoflakes. a&b) TEM and c&d) HRTEM images of a&c) WS$_2$ and b&d) BP nanoflakes. 2D BP/WS$_2$ hybrids present a cost-effective method for increasing the demand of solar light and are used as a noble metal-free NIR-motivated photocatalyst. Adapted with permission from Zhu et al. (2018), Copyright © 2017, Elsevier B.V. All rights reserved.

TEM (HRTEM) is used to analyze the crystallinity of the BP samples, and the existence of lattice fringes having the value of the fringe around 0.27 nm and 0.22 nm were supposed to correspond to the (100) and (002) planes value for WS$_2$ and BP nanosheet respectively (Figure 4.6c–d) (Wang et al. 2015; Du et al. 2015).

The size of the WS$_2$ structure has a great interaction with its assembly. It is noted that when compared to relatively large-sized WS$_2$ structures, the small-sized WS$_2$ nanoflakes have a great interaction for assembly on the top surface region of BP nanosheet and this might be due to the similar 2D structures of both BP as well as WS$_2$. The low-resolution image (LRTEM), TEM image, and HRTEM pictures of the as-prepared hybrids of BP/WS$_2$ material are shown in Figure 4.7, and Figure 4.7a–b show the morphology as similar to a small irregular sheet, with the presence of WS$_2$ nanoflakes on the nanoflakes' (BP) surface. HRTEM images taken from three different regions on the TEM image (white dotted circle) were selected (Figures 4.7c–f) from Figure 4.6b to confirm the BP/WS$_2$ hybrid components. In the first circle, only the presence of BP is confirmed by the lattice fringes having a value of d-spacing of 0.22 nm (corresponding to (002) plane) is prominent in the area of Figure 4.7d′, confirmed the presence of BP nanoflakes.

The presence of WS$_2$'s lattice fringes having a value of d-spacing approximately 0.27 nm (allocated to (100) plane) and 0.62 nm (allotted to (002) plane) was

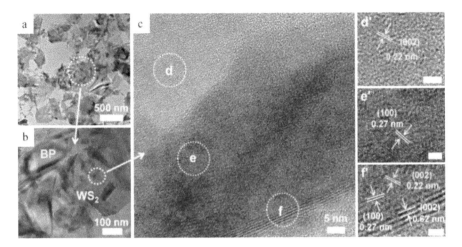

FIGURE 4.7 The internal morphology of the BP/WS$_2$ hybrid. a) Low-magnification TEM(LRTEM), b) TEM, and c–f,d′–f′) HRTEM pictures of hybrid BP/WS$_2$ material. The scale bar length in d′–f′) is 2 nm. Adapted with permission from Zhu et al. (2018), Copyright © 2017, Elsevier B.V. All rights reserved.

confirmed from the area of Figure 4.7e–f. Additionally, Figure 4.7f′ also confirmed the presence of some lattice fringes' spacing values consistent with BP nanoflakes. These results strongly indicated the presence of WS$_2$ nanoflakes on the BP surface (Zhu et al. 2018). BP and WS$_2$ individually can produce only a trace amount of H$_2$, while there is an enhancement of H$_2$ production when the combination of BP/ WS$_2$ hybrid is introduced and having an approach below 780 nm and 808 nm under the treatment of NIR laser light technique (Figure 4.8). After getting the excited state under NIR light, the as-prepared BP generates holes and electrons in the valance band (VB) and covalent band (CB), respectively, and the level of CB of BP is appropriate for the production of H$_2$ when BP is photoexcited. Despite this, BP displays an insignificant photocatalytic action on H$_2$ production due to rapid charge recombination. Conversely, due to an interaction of BP with WS$_2$, it is expected that the BP having the CB electrons are expeditiously transported to WS$_2$ nanoflakes in BP/WS$_2$ hybrid structure possibly due to the minor value of work function (5.57 eV) of WS$_2$ (Vovusha and Sanyal 2015). There is a reduction of H+ (water) for generating the H$_2$ after a lot of electrons are trapped by WS$_2$. Consequently, to contribute to a water reduction process to H$_2$ for the case of charge parting and encouraging the photogenerated electrons, more opportunities exist due to the presence of WS$_2$ to make the process effective as well as reliable.

WS$_2$ separates the charge carriers, works as an effective and reliable co-catalyst during the process of H$_2$ manufacture, and is useful for improving the production efficiency of H$_2$. Moreover, using photoelectrochemical-based properties, transient absorption spectroscopy, and electron reductive reactions, the efficient way for charge parting processes has been discussed in detail (Zhu et al. 2018).

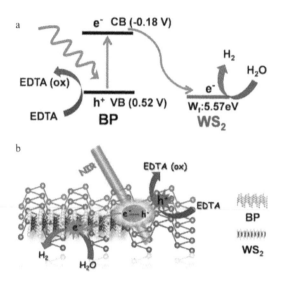

FIGURE 4.8 The process of NIR light-motivated photocatalytic-based H_2 creation. BP and WS_2 individually can produce only trace amount of H_2 while there is an enhancement of H_2 production when the combination of BP/ WS_2 hybrid is introduced and having an approach below 780 nm and 808 nm under the treatment of NIR laser light irradiation technique. Adapted with permission from Zhu et al. (2018), Copyright © 2017, Elsevier B.V. All rights reserved.

4.4.3 Heterostructures of Black Phosphorus with WSe₂

The junction formed as a result of two dissimilar semiconducting materials is known as a semiconductor heterojunction and such kinds of junctions are very important and play a key role in current semiconductor-based electronics, field-effect transistors, and optoelectronics as applied for logic circuits. Li et al. (2017) developed logic optoelectronic devices and gate-controlled logic rectifiers based on tungsten diselenide (WSe_2) and stacked black phosphorus (BP) heterojunctions.

The BP–WSe_2 heterojunction is grown with a 300-nm-thick thermal oxide-based film on the Silica glass sheet and the configuration of the device is schematically illustrated in Figure 4.9a. The SEM image (Figure 4.9b) demonstrates a color-decoration of a distinctive BP–WSe_2-heterojunction-based device mostly used to discriminate the device materials. In order to conclude about the width of every constituent of the composite film, AFM (Figure 4.9c) in a tapping style was engaged to measure the width value of both BP and WSe_2. The thickness of WSe_2 and the BP are 4.4 and 7.0 nm, respectively, representing that both of them are few-layer (FL) structures (Li, Yu, et al. 2014; Zhang et al. 2014).

The BP, the WSe_2, and the heterojunction region were further characterized by Raman spectroscopy. As shown in Figure 4.9d, the characteristic Raman spectrum of BP (Ag^1 peak at 365 /cm, B_2g peak at 440/cm, and Ag^2 peak at 470/cm) (Li, Yu, et al. 2014; Li et al. 2015) and WSe_2 (Ag^1 peak at 250/cm, E_2g^1 peak at 257/cm, and B_2g^1 peak at 307/cm) (Yu et al. 2015; Luo et al. 2013) are perceived in the consistent Raman

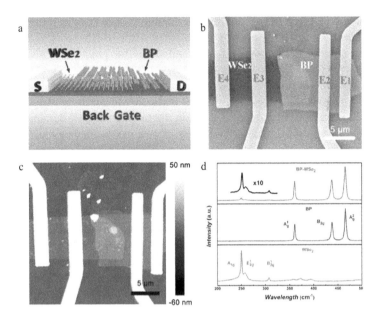

FIGURE 4.9 Development of logic optoelectronic devices and gate-controlled logic rectifiers based on tungsten diselenide (WSe$_2$): a) Schematically illustration of the heterojunction of BP–WSe$_2$ is grown with a 300-nm-thick thermal oxide film on Silica glass sheet and the configuration of the device. b) SEM image of a. demonstrates a color-decoration of a distinctive BP–WSe$_2$ based heterojunction device mostly used to discriminate the device materials. c) Surface morphology representation through AFM analysis d) The BP, the WSe$_2$, and the heterojunction region was further characterized by Raman spectroscopy. As mentioned in Figure 4.9d, typical Raman spectrum of BP (Ag^1 peak at 365/cm, B$_2$g peak at 440/cm, and Ag$_2$ peak at 470/cm) and WSe$_2$ (Ag^1 peak at 250 cm^{-1}, E$_2$g^1 peak at 257/cm, and B$_2$g^1 peak at 307/cm). Adapted with permission from Li et al. (2017) Copyright 2017 WILEY–VCH Verlag GmbH & Co. KGaA, Weinheim.

spectrum. Moreover, rectifying and photovoltaic properties, due to the ambipolar (gate-tunable) charge movers, mobility values in case of WSe$_2$ and BP resulted in a bendable, energetic, and comprehensive variation on the heterojunctions as an isotope (p–n) and isotope (p–p and n–n) diodes. This evidence strongly recommended the high-performance value of logic rectifiers and logic optoelectronic devices based on such characteristics developed from BP–WSe$_2$-based heterojunction diodes. A light signal can be converted to an electric one by logic optoelectronic devices through applied gate voltages (Li et al. 2017). In particular, these ideas might be supportive to extend the applications of 2D-based crystals for electronic devices.

4.5 SUMMARY

Proof of concept varieties BP-based heterostructures were demonstrated successfully, while significant efforts have been made to optimize performance. However, the same priority should be given to the stability and durability of devices in order to achieve practical applications. The extensive preparation of stable BPNSs remains a challenge, which also reduces the building of heterostructures for practical applications. As heterostructures based on the BPNS grow, new methods of scalable synthetic design are very important in order to preserve the properties of each material and manage the number of layers. It is believed that BP research will make the development of MDWs a lifelong hotspot that could assist in the finding of the killer application of BP, leading ultimately to significant advances in new nanotechnology and marketing of BP-based devices. Meanwhile, research into BP-based heterostructures can be a reference point for other novel 2D crystals beyond BP and can be used as a general performance improvement strategy for BP-based van der Waals (vdWs) devices.

4.6 ACKNOWLEDGMENTS

The research was partially supported by financial support from the Science and Technology Development Fund (Nos. 007/2017/A1 and 132/2017/A3), Macao Special Administration Region (SAR), China, and National Natural Science Fund (Grant Nos. 61875138, 61435010, 61805147, and 6181101252), and the Science and Technology Innovation Commission of Shenzhen (KQTD2015 0324162703, JCYJ20150625103619275, JCYJ20180305125141661, and JCYJ2017 0811093453105). The authors also acknowledge the support from the Instrumental Analysis Center of Shenzhen University (Xili Campus).

REFERENCES

Ansari, Sajid Ali, Ziyauddin Khan, Mohammad Omaish Ansari, and Moo Hwan Cho. 2016. "Earth-abundant stable elemental semiconductor red phosphorus-based hybrids for environmental remediation and energy storage applications." *RSC Advances* 6(50): 44616–44629.

Chen, Jian-Hao, Chaun Jang, Shudong Xiao, Masa Ishigami, and Michael S. Fuhrer. 2008. "Intrinsic and extrinsic performance limits of graphene devices on SiO2." *Nature Nanotechnology* 3(4): 206–209.

Chen, Peng, Jianyong Xiang, Hua Yu, Guibai Xie, Shuang Wu, Xiaobo Lu, Guole Wang, Jing Zhao, Fusheng Wen, and Zhongyuan Liu. 2015. "Gate tunable MoS2–black phosphorus heterojunction devices." *2D Materials* 2(3): 034009.

Choi, Min Sup, Gwan-Hyoung Lee, Young-Jun Yu, Dae-Yeong Lee, Seung Hwan Lee, Philip Kim, James Hone, and Won Jong Yoo. 2013. "Controlled charge trapping by molybdenum disulphide and graphene in ultrathin heterostructured memory devices." *Nature Communications* 4(1): 1–7.

Deng, Yexin, Zhe Luo, Nathan J. Conrad, Han Liu, Yongji Gong, Sina Najmaei, Pulickel M. Ajayan, Jun Lou, Xianfan Xu, and Peide D. Ye. 2014. "Black phosphorus–monolayer MoS2 van der Waals heterojunction p–n diode." *ACS Nano* 8(8): 8292–8299.

Du, Yichen, Xiaoshu Zhu, Ling Si, Yafei Li, Xiaosi Zhou, and Jianchun Bao. 2015. "Improving the anode performance of WS2 through a self-assembled double carbon coating." *The Journal of Physical Chemistry C* 119(28): 15874–15881.

Geim, A. K., and I. V. Grigorieva. 2013. "Van der Waals heterostructures." *Nature* 499: 419.

Hu, Zhuofeng, Zhurui Shen, and C. Yu Jimmy. 2017. "Phosphorus containing materials for photocatalytic hydrogen evolution." *Green Chemistry* 19(3): 588–613.

Huang, Le, Nengjie Huo, Yan Li, Hui Chen, Juehan Yang, Zhongming Wei, Jingbo Li, and Shu-Shen Li. 2015. "Electric-field tunable band offsets in black phosphorus and MoS2 van der Waals pn heterostructure." *The Journal of Physical Chemistry Letters* 6(13): 2483–2488.

Jiang, Yaqin, Lili Miao, Guobao Jiang, Yu Chen, Xiang Qi, Xiao-fang Jiang, Han Zhang, and Shuangchun Wen. 2015. "Broadband and enhanced nonlinear optical response of MoS 2/graphene nanocomposites for ultrafast photonics applications." *Scientific Reports* 5: 16372.

Levendorf, Mark P., Cheol-Joo Kim, Lola Brown, Pinshane Y. Huang, Robin W. Havener, David A. Muller, and Jiwoong Park. 2012. "Graphene and boron nitride lateral heterostructures for atomically thin circuitry." *Nature* 488(7413): 627–632.

Li, Diao, Antonio Esau Del Rio Castillo, Henri Jussila, Guojun Ye, Zhaoyu Ren, Jintao Bai, Xianhui Chen, Harri Lipsanen, Zhipei Sun, and Francesco Bonaccorso. 2016. "Black phosphorus polycarbonate polymer composite for pulsed fibre lasers." *Applied Materials Today* 4: 17–23.

Li, Dong, Xiaojuan Wang, Qichong Zhang, Liping Zou, Xiangfan Xu, and Zengxing Zhang. 2015. "Nonvolatile floating-gate memories based on stacked black phosphorus–boron nitride–MoS2 heterostructures." *Advanced Functional Materials* 25(47): 7360–7365.

Li, Dong, Biao Wang, Mingyuan Chen, Jun Zhou, and Zengxing Zhang. 2017. "Gate-controlled BP–WSe2 heterojunction diode for logic rectifiers and logic optoelectronics." *Small* 13(21): 1603726.Li, Hong-Ju, Ling-Ling Wang, Han Zhang, Zhen-Rong Huang, Bin Sun, Xiang Zhai, and Shuang-Chun Wen. 2014. "Graphene-based mid-infrared, tunable, electrically controlled plasmonic filter." *Applied Physics Express* 7(2): 024301.

Li, Likai, Yijun Yu, Guo Jun Ye, Qingqin Ge, Xuedong Ou, Hua Wu, Donglai Feng, Xian Hui Chen, and Yuanbo Zhang. 2014. "Black phosphorus field-effect transistors." *Nature Nanotechnology* 9: 372–377.

Liu, Hanwen, Li Tao, Yiqiong Zhang, Chao Xie, Peng Zhou, Hongbo Liu, Ru Chen, and Shuangyin Wang. 2017. "Bridging covalently functionalized black phosphorus on graphene for high-performance sodium-ion battery." *ACS Applied Materials & Interfaces* 9(42): 36849–36856.

Liu, Zheng, Lulu Ma, Gang Shi, Wu Zhou, Yongji Gong, Sidong Lei, Xuebei Yang, Jiangnan Zhang, Jingjiang Yu, and Ken P. Hackenberg. 2013. "In-plane heterostructures of graphene and hexagonal boron nitride with controlled domain sizes." *Nature Nanotechnology* 8(2): 119–124.

Luo, Shaojuan, Jinlai Zhao, Jifei Zou, Zhiliang He, Changwen Xu, Fuwei Liu, Yang Huang, Lei Dong, Lei Wang, and Han Zhang. 2018. "Self-standing polypyrrole/black phosphorus laminated film: promising electrode for flexible supercapacitor with enhanced capacitance and cycling stability." *ACS Applied Materials & Interfaces* 10(4): 3538–3548.

Luo, Xin, Yanyuan Zhao, Jun Zhang, Minglin Toh, Christian Kloc, Qihua Xiong, and Su Ying Quek. 2013. "Effects of lower symmetry and dimensionality on Raman spectra in two-dimensional WSe 2." *Physical Review B* 88(19): 195313.

Mu, Haoran, Shenghuang Lin, Zhongchi Wang, Si Xiao, Pengfei Li, Yao Chen, Han Zhang, Haifeng Bao, Shu Ping Lau, and Chunxu Pan. 2015. "Black phosphorus–polymer composites for pulsed lasers." *Advanced Optical Materials* 3(10): 1447–1453.

Pomerantseva, Ekaterina, and Yury Gogotsi. 2017. "Two-dimensional heterostructures for energy storage." *Nature Energy* 2(7): 1–6.

Shang, Meng, Wenzhong Wang, Ling Zhang, Songmei Sun, Lu Wang, and Lin Zhou. 2009. "3D Bi2WO6/TiO2 hierarchical heterostructure: controllable synthesis and enhanced visible photocatalytic degradation performances." *The Journal of Physical Chemistry C* 113(33): 14727–14731.

Tian, Xiao-Qing, Lin Liu, Zhi-Rui Gong, Yu Du, Juan Gu, Boris I. Yakobson, and Jian-Bin Xu. 2016. "Unusual electronic and magnetic properties of lateral phosphorene–WSe2 heterostructures." *Journal of Materials Chemistry C* 4(27): 6657–6665.

Vovusha, Hakkim, and Biplab Sanyal. 2015. "Adsorption of nucleobases on 2D transition-metal dichalcogenides and graphene sheet: a first principles density functional theory study." *RSC Advances* 5(83): 67427–67434.

Wang, Hui, Xianzhu Yang, Wei Shao, Shichuan Chen, Junfeng Xie, Xiaodong Zhang, Jun Wang, and Yi Xie. 2015. "Ultrathin black phosphorus nanosheets for efficient singlet oxygen generation." *Journal of the American Chemical Society* 137(35): 11376–11382.

Wang, Xin, Han Li, Hui Li, Shuai Lin, Wei Ding, Xiaoguang Zhu, Zhigao Sheng, Hai Wang, Xuebin Zhu, and Yuping Sun. 2020. "2D/2D 1T-MoS2/Ti3C2 MXene heterostructure with excellent supercapacitor performance." *Advanced Functional Materials* 30(15): 0190302.

Wang, Xinchen, Kazuhiko Maeda, Xiufang Chen, Kazuhiro Takanabe, Kazunari Domen, Yidong Hou, Xianzhi Fu, and Markus Antonietti. 2009. "Polymer semiconductors for artificial photosynthesis: hydrogen evolution by mesoporous graphitic carbon nitride with visible light." *Journal of the American Chemical Society* 131(5): 1680–1681.

Xu, Yanhua, Wenxi Wang, Yanqi Ge, Hanyu Guo, Xiaojian Zhang, Si Chen, Yonghong Deng, Zhouguang Lu, and Han Zhang. 2017. "Stabilization of black phosphorous quantum dots in PMMA nanofiber film and broadband nonlinear optics and ultrafast photonics application." *Advanced Functional Materials* 27(32): 1702437.

Yu, Xiaoyun, Mathieu S. Prévot, Néstor Guijarro, and Kevin Sivula. 2015. "Self-assembled 2D WSe 2 thin films for photoelectrochemical hydrogen production." *Nature Communications* 6: 7596.

Zhang, Chenzhen, Nasir Mahmood, Han Yin, Fei Liu, and Yanglong Hou. 2013. "Synthesis of phosphorus-doped graphene and its multifunctional applications for oxygen reduction reaction and lithium ion batteries." *Advanced Materials* 25(35): 4932–4937.

Zhang, Guigang, Zhi-An Lan, and Xinchen Wang. 2016. "Conjugated polymers: catalysts for photocatalytic hydrogen evolution." *Angewandte Chemie International Edition* 55(51): 15712–15727.

Zhang, Y. J., T. Oka, Ryo Suzuki, J. T. Ye, and Y. Iwasa. 2014. "Electrically switchable chiral light-emitting transistor." *Science* 344(6185): 725–728.

Zhu, Mingshan, Chunyang Zhai, Mamoru Fujitsuka, and Tetsuro Majima. 2018. "Noble metal-free near-infrared-driven photocatalyst for hydrogen production based on 2D hybrid of black Phosphorus/WS2." *Applied Catalysis B: Environmental* 221: 645–651.

5 3D Structures Based on 2D BP

5.1 WHY 3D STRUCTURES OF BP ARE IMPORTANT

After using the 2D black phosphorous (BP) nanosheets as the basic unit for the building of a 3D architecture, geometries is considered as an operative and valuable method to represent the properties of 2D structures of materials (Yun et al. 2018; Wang et al. 2018; Choi, Yang, et al. 2012; Zhang, Li, and Lou 2017; Zhang, Wang, et al. 2018). Comparatively, 3D systems have a lot of advantages over 2D structures, including the following: 1) the 3D architectures are provide some suitable number of functional sites for reaction and ion absorption due to restocking of 2D nanosheets having large surface energy; 2) 3D structures of BP usually afford suitable pores or channels abundant to enable charge transport and ion dispersion; and 3) for electron transport, the abundant paths transforming 2D nanosheets to 3D structures form an interconnected scaffold. The up-grading of 2D material and the resultant 3D material structures have great technological, scientific, and commercial significance among the materials having a vital role for the well-organized energy storage and conversion due to the transport kinetics of electrons and ions (Wen et al. 2019). This chapter discusses the fabrication process of BP-based 3D structures and their morphological engineering using different deposition techniques. A comparative study of BP with other 3D structures of layered material is organized, which differentiates BP as a unique candidate for potential and biomedical applications.

5.2 MORPHOLOGICAL ENGINEERING OF 3D ARCHITECTURE OF BP

5.2.1 HONEYCOMB-LANTERN-INSPIRED 3D STRUCTURES OF 2D BLACK PHOSPHORUS

During the past decade, 2D shapes supercapacitors with the design of planar stretchable electrodes—named as a bridge-island electrode (Kim et al. 2013), textile electrode (Hu et al. 2010; Xu et al. 2015), and wrinkle-like electrode (Yu et al. 2016; Qi, Liu, et al. 2015; Yu et al. 2009; Chen et al. 2013; Niu et al. 2013)—are mainly reported (Figure 5.1a). In the case of wearable devices, there is a deficiency in the utilization of the whole 3D space, which might be due to the limited 2D surface of 2D

DOI: 10.1201/9781003217145-5

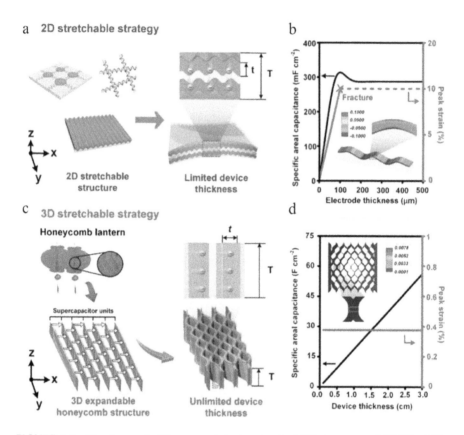

FIGURE 5.1 Development of honeycomb lantern-inspired 3D structures of 2D BP. a) The thickness of device (T) depends on the construction of stretchy and flexible planar conductors, and the efficiency of the 2D stretchable supercapacitor strongly depends upon enhancing the thickness (t) value of the electrode for the developed value of specific areal capacitance. b) The finite element modeling analysis due to thicker electrodes (formed as a result of prolonged the thickness of device due to higher mass loading) that suffer longer ionic transport paths; as a result of increasing the electrode thickness the peak strains value of stretchable and flexible electrodes are improved, which limits the stretch-ability and flexibility of the supercapacitors. c) Supercapacitors are extended to 3D expandable and bendable honeycomb forms (having infinite value of thickness for device and enhancing the specific areal capacitance value). d) The supercapacitor units and the value of strain is scattering in the perpendicular direction (z-axis) in the bendable and stretchy honeycomb crystal structure practice an out-of-plane (oop) twisting are recognized during the stretching process. Adapted with permission from Lv et al. (2018), Copyright © 2018, WILEY–VCH Verlag GmbH & Co. KGaA, Weinheim.

stretchable supercapacitors. Consequently, to overtake the problem of 2D stretchable and flexible supercapacitors, it is necessary to develop 3D stretchable supercapacitors with high mass loading efficiency to improve the specific areal capacitance value. The specific areal capacitance and internal resistance are increased in conventional 2D stretchable supercapacitors. This is due to thicker electrodes (formed as a result of prolonged the thickness of device due to higher mass loading) that suffer longer ionic transport paths, and as a result of increasing the electrode thickness the peak strains value of stretchable and flexible electrodes are improved, which limits the stretch-ability and flexibility of the supercapacitors (Figure 5.1b) (Xie and Wei 2014; Rogers, Someya, and Huang 2010). Lv et al. (2018) developed highly flexible polypyrrole/black phosphorus oxide-carbon nanotube (PPy/BPO-CNT) film electrodes, which combined with the expandable honeycomb structure to create a powerful extension in mechanical and electrochemical reliability for the 3D bendable and stretchable and flexible supercapacitors. The basic construction of honeycomb lantern is simple and easily made by adjusting a piece of paper with adhesives (to fix the presence of piece of paper) into a 3D stretchable and bendable artifact model, with an internally expandable honeycomb structure (Figure 5.1c).

Additionally, for the formation of 3D stretchable supercapacitors (for the deposition of porous PPy/BPO composite electrodes on the surface of carbon nanotube (CNT) films, the electrode position technique was used. The supercapacitor units and the strain are scattering in the perpendicular track (z-axis) in the bendable and stretchy honeycomb arrangement; an out-of-plane (oop) winding is recognized during the stretching process (Figure 5.1d) (Lv et al. 2018). Additionally, for ion transport and effective stress relief, such electrodes exhibit high porosity and flexibility and the designed electrodes were further connected with expandable honeycomb structures to form 3D supercapacitors.

The resultant supercapacitors (with stretch-ability and device-thickness-independent ion transport path) are formed due to the combination of the flexible and stretchable PPy-based composite electrodes with unique expandable honeycomb structure (Luo et al. 2018). Moreover, as a result of the deformation process, the electrochemical performance is maintained by PPy/BPO-CNT, a hierarchical and porous structure of composite electrodes; meanwhile, for higher areal energy storage the stretch-ability of expandable honeycomb structure and ion-transport path of device-thickness facilitate the (thicker, stretchable and more bendable) supercapacitors to be modified into anticipated 3D architecture (Zhu et al. 2017; Chen, Cai, et al. 2017; Zhang, Liao, et al. 2018).

5.2.2 3D-Printed Scaffolds from 2D Black Phosphorus

The well-defined design and construction of biomaterials may support to temporospatially regulate biophysical and biochemical approaches (Li, Xiao, and Liu 2017). 3D printing, proposals great projections of various applications in the fields such as cell and tissue engineering and is regarded as an innovator in engineering technology (Baumann et al. 2017; Pawlyn 2013; Murphy and Atala 2014). For the preparation of 3D-printed scaffolds, the BP nanosheets were synthesized via a simple liquid exfoliation method. Around 80 mL N methyl pyrrolidone (NMP) solvent was used for the dispersion of 30 mg of BP crystal powder, the resultant mixture was sonicated in an ice bath for 24 hours and the supernatant containing BP nanosheets was decanted gently after the residual unexfoliated crystal powder was removed. Yang et al. (Yang, Yin, et al. 2018) developed a BP-reinforced 3D-printed scaffold (via a simple surface modification) as a result of mixing BP nanosheets into

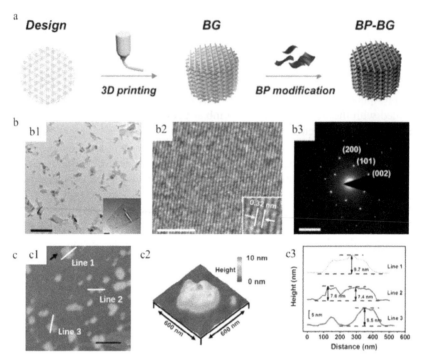

FIGURE 5.2   3D-printed scaffolds from 2D black phosphorus. a) The schematic representation of 3D printing of BP-BG scaffolds. b) TEM images of BP nanosheets show the dispersion of nanosheets as well as internal morphology. b1) TEM image of highly dispersed nanosheets of BP with a scale bar of 500 nm and characteristic nanoflake-like structure (inset) with a scale bar of 100 nm. b2) High-resolution TEM (HRTEM) picture showing the internal morphology of BP nanosheet with a scale bar of 5 nm. b3) SAED of BP nanosheet with a scale bar of 5 nm shows the crystallinity of BP nanosheets. c) Atomic force microscopy (AFM) analysis of BP nanosheets used to calculate the thickness of the BP sheet as well to monitor the surface morphology and dispersion of BP nanosheets. c1) The AFM image of highly dispersed BP nanosheets with a scale bar of 1 μm. c2) 3D reconstruction of a representative nanosheet of BP in c3). The color bar refers to the relative height distributions on the surface of the nanosheet of BP. Adapted with permission from Yang, Yin, et al. (2018), Copyright © 2018, WILEY–VCH Verlag GmbH & Co. KGaA, Weinheim.

the bio-glass (BG) scaffolds for the preparation of a bifunctional BP-BG scaffold (Figure 5.2a) and dry obtained BG scaffolds were used for nano characterizations. Transmission electron microscope (TEM) (Figure 5.2b1), was used to check the size and shape of 2D BP nanosheets at micro-scale, demonstrating that these size-controlled (~500 nm) nanosheets were free-standing. The presence of lattice fringes with a value of d spacing around 0.32 nm and the Specific area electron diffraction pattern (SAED) is shown in Figure 5.2b2 to support the parent crystal structure's orthogonal type structure and crystallinity of BP (Figure 5.2b3) (Yasaei et al. 2015; Castellanos-Gomez 2016). The value of thickness ascertained by the topographic morphology and cross-sectional analysis of the BP nanosheets is less than 10 nm (Figure 5.2c1–c3) as shown in the atomic force microscope (AFM) image, indicating the surface morphology and the successful synthesis of ultrathin BP nanosheets. The 3D printing technique is used for the fabrication of BG scaffolds and comparatively BP-BG scaffolds with a 3D geometrical structure are more effective and reliable than the 2D geometrical topographical substrates and facilitate the process of cell, providing adhesion in a more effective way (Zaman et al. 2006; Fraley et al. 2010; Broussard et al. 2015).

5.2.3 BLACK PHOSPHORUS SPONGES BY THE MODIFIED ELECTROCHEMICAL APPROACH

The combination of 3D sponge with ultrathin BP nanosheets was prepared as a result of the electrochemical approach, and during the synthesis process, the protection for oxidation process of BP is carried out using a bromide electrolyte ion. Wen et al. (2019) used the ambient atmosphere and developed a simple and rapid electrochemical technique and synthesized 3D BP sponges. The resultant prepared sponge consists of a BP nanosheet with a thickness of less than 4.0 nm and within the size range (micrometer range). Under ambient conditions, a two-electrode system such as Pt sheet as the counter electrode and bulk BP crystal as the cathode was introduced for the synthesis of the BP sponges, while tetrabutylphosphonium bromide (Tbb) resolved in DMF was used as an electrolyte, as shown in Figure 5.3a. The basic construction of the electrochemical circuit is very simple and reliable. For the separation process of two electrodes such as the bulk BP cathode with the Pt anode, the electrochemical circuit is linked with an H-type electrolytic type cell. The electrolytic cell has a portion of the cation-interchange film in the center of the connection, while between the two

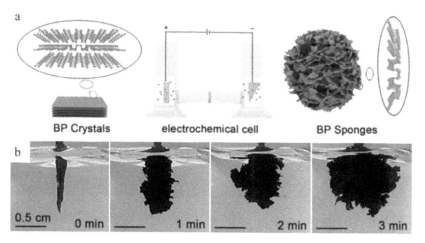

FIGURE 5.3 The synthesis process of BP based sponge: a) Schematic representation of bulk crystal of BP including the basic development of electrochemical process cell and the appearance of a BP sponge-like structure. b) The photograph was engaged during a different time interval of the synthesis process. Adapted with permission from Royal Society of Chemistry (Wen et al. 2019).

electrodes the tetrabutylphosphonium cations can diffuse. The BP sponge is prepared with a continuous functional bias voltage of around 5 V, as shown in Figure 5.3b, and sponge-like morphology of BP structure is observed. The surface morphology (the BP sponge contains huge BP nanosheets with a size range in micrometers) of synthesized BP sponge analyzed by scanning electron microscope (SEM) is shown in Figure 5.4a–b. Other nanoscale characterizations such as TEM, AFM, and SAED are further used to discover the basic unit structure in the semi-connected BP sponge.

AFM is used to investigate the thickness of the ultrathin BP nanosheets as shown in Figure 5.4c, the BP layers having the thickness value ranging 1.8 to 4.0 nm. It is strongly recommended that the BP-based sponge is collected as a result of FL- BP nanosheets (Figure 5.4d). The crystal structure and the size and shape of the BP units, the thickness and internal morphology, are auxiliary scrutinized by TEM and the corresponding SAED pattern (Yasaei et al. 2015; Hultgren, Gingrich, and Warren 1935). The forbidden reflections such as (010) and (110) confirm the ultrathin BP.

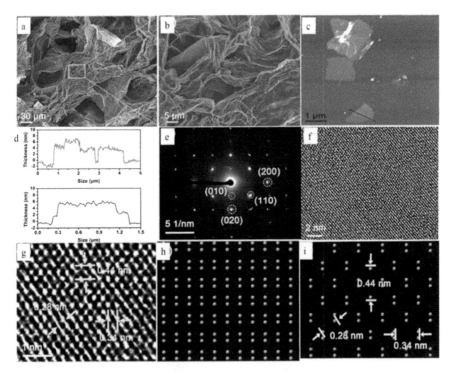

FIGURE 5.4   Characterization of BP sponge by the modified electrochemical approach. a&b) SEM images shows the surface morphology of BP sheets. c) AFM image shows the surface morphology of nanosheets in BP sponge. d) Height profiles of nanosheets from calculated from image c. e) SAED pattern shows the crystallinity of BP nanosheet. f) TEM images show the internal morphology of BP based nanosheets. g) Highly magnified HR-TEM image of the structures of BP nanosheets. h) Multilayer and i) single-layer type BP. Adapted with permission from Royal Society of Chemistry (Wen et al. 2019).

The single-layer BP is confirmed from (010) reflection and the existence of single-layer (SL) BP nanosheets is confirmed from the high-intensity (110) reflection as mentioned by the simulated SAED arrangements (Figure 5.4e).

Figure 5.4f–g illustrates the lattice fringes value as 0.28, 0.34, and 0.44 nm and such kind of atomic organization matches that of SL-BP but is unlike that of multi-layered BP (ML-BP) (Figure 5.4h–i) (Hultgren, Gingrich, and Warren 1935). These results strongly indicate that the formation of semi-linked BP sponge confirms the presence of ultrathin BP nanosheets.

5.2.4 GRAPHENE OXIDE/BLACK PHOSPHORUS 3D NANOFLAKES AEROGEL

The use of graphene oxide (GO) has proven to be operative and reliable in photothermal therapy (PTT) (Zhang, Guo, et al. 2011; Li et al. 2012; Yang et al. 2010, 2012; Liu et al. 2013) and BP have also been found as a favorable candidate in PTT against tumor cell growth (Sun et al. 2016; Sun et al. 2015). Xing et al. (2017) reported for the first time a different kind of hybrid three-dimensional (3D) aerogel as a combination of black phosphorus nanoflakes (BPNFs) and graphene oxide (GO). Figure 5.5a shows the as-prepared BPNFs have a size in the range of several hundred nanometers and the lattice fringes of 2.24 Å allotted to (014) plane 41 from the SAED pattern images (Figure 5.5b). These pieces of evidence strongly suggest the presence and maintenance of BP crystal structures during the preparation process. Figure 5.5c–d shows the atomic force microscope (AFM) for BPNFs having a thickness in the ranging from 10~30 nm for BPNFs and on the other side, the GO nanosheets have greater dimensions (with several micrometers) and a smaller thickness value around 1.5 nm (Figure 5.5e–f). Additionally, GO nanosheets covered the BPNFs in these aerogels and showed robust photothermal stability under ambient conditions without any protection. The gelation formed as a result of hybrid GO with BP. The GO caused the protection of BP from degradation in air and increased the photothermal characteristics of GO, which might provide a new direction to improve the BP-GO-based hybrid materials for biological applications. Finally, it was concluded that as-prepared composite GO/BPNFs aerogels are responsible for significantly boosted photothermal (Figure 5.5g–h) and electrical properties (Figure 5.5i) equated to that of neat GO-based aerogel as a result of incorporation of BP (Xing et al. 2017).

5.2.5 CELLULOSE/BP NANOSHEETS (BPNSS)-BASED HYDROGEL POSSESSES 3D NETWORKS

Cellulose is a biopolymer, found as a rich source on the earth's crust, and has a wide range of exceptional properties, such as outstanding biocompatibility, biodegradability, large polarity, no toxicity, and transparency. The composite of cellulose hydrogels having functionalization with an inorganic quantum dot have revealed significant potential in the field of biomedicines among all the cellulose-based composites (Chang and Zhang 2011) due to some advantages. 1) Water-filled cellulose hydrogels are totally behaving as real green materials capable of biocompatibility within cells/tissues. As a result of the degradation process inside the living body system, the water, carbon dioxide, and other required ingredients are obtained. 2) As a

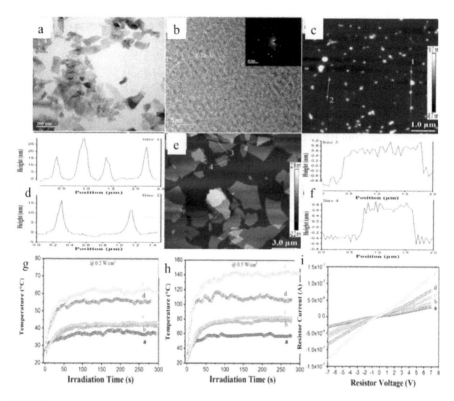

FIGURE 5.5 Morphology characterization of black phosphorus nanoflakes (BPNFs) and graphene oxide (GO) nanosheets used for the preparation of 3D GO/BPNF-based composite aerogels. a) TEM pictures showing the dispersion as well as internal morphologies and structure of BPNFs. b) The lattice fringes value of 2.24 Å allocated to (014) plane 41 from the SAED images. c&d) The AFM images for BPNFs having a thickness ranging from 10~30 nm for synthesized BPNFs. e&f) The GO-based nanosheets show the dimensions (with several micrometers) and with the reduced value of thickness in the range of 1.5 nm. g&h) Showing that as-prepared composite GO/BPNFs aerogels are responsible for significantly boosted photothermal equated to that of neat GO-based aerogel a: 0 wt% (i.e. bare/simple GO aerogel), and increasing the BP contents such as b: 7.5 wt%, c: 9.6 wt%, d: 13.4 wt% and e: 26.3 wt%, respectively. i) Electrical-based properties of composite GO/BPNFs aerogels equated to that of neat GO aerogel as a result of the incorporation of BP (same BP and GO contents as in case of g&h) (Xing et al. 2017).

result of the addition of nanoparticles (NPs) into cellulose hydrogel matrices, there is an improvement in stress and strain, and nanoparticles share photoluminescence, conductivity catalytic, and magnetic properties to the hydrogel depending on the properties of nanoparticles (Chang and Zhang 2011).

For the synthesis of cellulose/BP nanosheets (BPNSs)-based hydrogel, the synthetic protocol is developed by Xing et al. (2018) and suggest three different steps (Figure 5.6) for the synthesis process. Step I involved the resolution of cellulose in

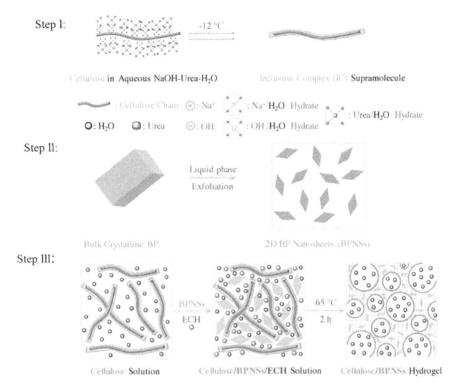

FIGURE 5.6 Synthetic protocol for cellulose/BP nanosheets (BPNSs)-based hydrogel. Step I: The resolution of cellulose in an aqueous-based media and basic NaOH 7 wt% –urea 12 wt% –H_2O 81 wt% system low temperature value of around –12°C. Step II: Liquid-phase exfoliation process (NMP used as a solvent) was conducted for obtaining BPNSs. Step III: Preparation process of BPNS-combined cellulose-based hydrogels. Adapted with permission from Xing et al. (2018), Copyright © 2018, WILEY–VCH Verlag GmbH & Co. KGaA, Weinheim.

sedimentary state basic NaOH 7 wt% –urea 12 wt% –H_2O 81 wt% system at −12°C. Cellulose chains can be resolved and separated in water due to the presence of three hydrates [Na+/H_2O (Na+-based hydrate), OH−/H_2O (OH-based hydrate), and urea/ H_2O (urea-based hydrate)]. In both intra-chain and inter-chains, all these hydrates are responsible for the destruction of an old huge network of hydrogen bonding of cellulose, and similarly, new ones are formed between cellulose chains and the three mentioned hydrates, as a result of keeping a low-temperature range (Step I, left). This unique interaction might help the rapid dissolution of cellulose chains, and the ending cellulose chains dispersed in water happens in the form of a network (right, Step I). In Step II, liquid-phase exfoliation process NMP used as a solvent was conducted for obtaining BPNSs. Additionally, in Step III, cross-linker of epichlorohydrin (ECH) was used in gelating cellulose chains in the occurrence of BPNSs in water, as a result, the synthesis process of BPNS-combined cellulose (cellulose/BPNSs) hydrogels is achieved (Xing et al. 2018).

Figure 5.7a shows the as-prepared BPNSs (in Step II) with a measurement ranging from 50 to 430 nm, and their high-resolution transmission electron microscopy (TEM) image is shown in Figure 5.7b. The crystalline lattice fringes of 0.323 and 0.223 nm allotted to the (012) and (014) planes of BP are clear from the TEM image. Besides, the crystalline structure of the BPNSs throughout the exfoliation process is confirmed as a result of using the flourier transform photograph (ptp) (Figure 5.7c) and the consistent SAED pattern (Figure 5.7d).

AFM was used to calculate the thickness of the as-prepared BPNSs as shown in Figure 5.7e and it is analyzed that BPNSs have a wide range of thickness from 5.1 to 10.8 nm (Figure 5.7f). The cellulose/BPNS composite hydrogels (Figure 5.7h) are obtained as a result of heating the uniform cellulose/solution with ECH (Figure 5.7g) at 65°C for 1.5 hours. On adding excess diluted H_2SO_4 solution, the redevelopment procedure of composite cellulose/BPNS composite hydrogels was successfully finished as shown in Figure 5.7i and the dialysis process is introduced in water for four days for cleaning and removing the inorganic salts (such as Na_2SO_4, NaOH, urea, and H_2SO_4) and excess ECH molecules (Figure 5.7j–k). In particular, during the long dialysis process, conditions such as dark conditions and low-temperature suppressed the degradation process of the BPNSs in these hydrogels. The resultant cellulose-based hydrogels have excellent flexibility and elasticity (Figure 5.7l) and possessed a special macroscopic shape similar to the formation of the vessels used (Sun et al. 2015; Guo et al. 2015).

Figure 5.8 shows the physical properties and typical morphology of cellulose hydrogels both with and without BPNSs. Figure 5.8a1 shows the typical 3D networks having some kind of unbalanced, porous-like structures (approximately the diameter of pore is ≈50–300 μm) of neat cellulose hydrogel without BPNSs; these pores provide a chance for the absorption and preservation of a large amount of water. The development of pores in cellulose-based composite hydrogels is possible due to the cross-linking process among poor and rich-cellulose regions. Consequently, a stable 3D network is formed after thin cellulose pore walls (Figure 5.8a2–a3).

To check the effect on the morphology of cellulose hydrogel of the presence of BPNSs, a controlled experiment was conducted and it was concluded that a similar kind of 3D networks are detected for the BP-combined cellulose-based composite hydrogels as shown in Figure 5.8b1–b3, as is the case without BPNSs. This evidence strongly recommended that for the formation of 3D networks structure of cellulose-based hydrogels, the accumulation of BPNSs was unable to create a significant effect on the phase separation performances. In summary, the achieved cellulose functionalized BPNS composite-based hydrogels were a combination of cellulose and BPNSs, which are anticipated to display unique biocompatibility and flexibility (Xing et al. 2018).

5.2.6 PALLADIUM NANOPARTICLES AND BLACK PHOSPHORUS HYBRIDS 3D STRUCTURE

There is a strong metal-support interaction (MSI) found among TiO_2 and Pd and the outstanding electronic properties of BP, and innovative anatase titanium dioxide nanosheets-BP (ATN-BP) based hybrids are prepared. Their backings for PdNPs are

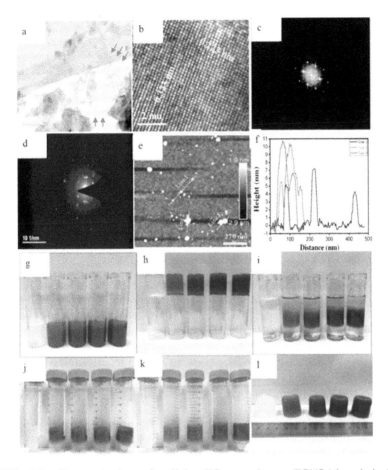

FIGURE 5.7 Characterization of cellulose/BP nanosheets (BPNSs)-based hydrogel possessing 3D networks. a) Transmission electron microscopy (TEM) for checking the dispersion and internal morphology of black phosphorus nanosheet (BPNSs). b) The crystalline lattice spacing of BPNSs is displayed by high-resolution (HR-TEM) image. c) The crystalline structures of the BPNSs throughout the exfoliation process are confirmed as a result of using the flourier transform photograph (ptp). d) The SEAD pattern shows the crystallinity of BPNSs. e) AFM images show the surface morphology of BPNSs. f) The consistent thickness/height profiles of BPNSs have a wide range of thicknesses from 5.1–10.8 nm from e). g) During different preparation stages the snapshot of cellulose/BPNS composite hydrogels having different quantities of BP 0 ppm, 95ppm, 190ppm, 285ppm, and 380 ppm (from the left to the right), respectively. h) In the existence of BPNSs in basic and aqueous (NaOH–urea–H_2O) system, the gelation of cellulose due to the support of cross-linker of epichlorohydrin (ECH). i) After the addition of dilute H_2SO_4 solution (5 wt%) there is a regeneration process of cellulose-based hydrogels and after that the dialysis procedure of hydrogels (cellulose-based) in water as a universal solvent j) before and k) after four days of storing time at low temperature in a cooling state (refrigerator) at low temperature 0–4°C. l) Cylinder shape (radius: 11 mm; height: 24 mm) formed as a resultant of cellulose-based hydrogels. Adapted with permission from Xing et al. (2018), Copyright © 2018, WILEY–VCH Verlag GmbH & Co. KGaA, Weinheim.

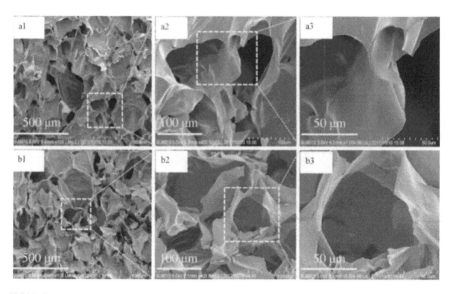

FIGURE 5.8 The physical properties and typical morphology of cellulose hydrogels both cases with and without BPNSs. a) The typical 3D networks having some kind of unbalanced type porous hybrid structures material (approximately pore diameter around ≈50–300 μm) of neat cellulose hydrogel without BPNSs, with a chance for absorption and preservation of a large amount of water due to these large pores. b1–b3) 3D cellulose functionalized BPNS composite hydrogel having a BP with small content with a concentration of 380 ppm. Adapted with permission from Xing et al. (2018), Copyright © 2018, WILEY–VCH Verlag GmbH & Co. KGaA, Weinheim.

recycled in the ethanol oxidation reaction (EOR). Wu et al. (2018) used the ball-milling method for the preparation of a hybrid of BP nanosheets and ATN under the influence of an argon gas atmosphere for 48 hours. As a result of the ball-milling method, the ATN crystal lattice suffers simple plastic distortion as a result of producing stresses and strains during the process (Shifu et al. 2005). Additionally, during the fabrication process, the cracking of the ATN is responsible for making several lattice flaws inside TiO_2.

The ATN-BP-based hybrids porous material was designed during the ball-milling process, which might be possible due to the unstable BP nanoflakes and ATN reaction. ATN-BP hybrids' porosity and the original surface areas of BP and ATN are checked using the Brunauer–Emmett–Teller (BET) technique.

It is found that separately the surface area of BP nanoflakes and ATN is around 6.3 and 83.5 m^2/ g, respectively and the combined ATN-BP-based hybrid material shows a BET surface area of 12.8 m^2/ g having a 3D structure after the reaction (Figure 5.9). After the distribution of ATN into minor catalectic, there is a process of developing strong P-O-Ti bonds. After the connection of small divided pieces of ATN connected with BP nanoflakes during the ball-milling process, it is possible. Moreover, phosphorus-doped ATN is formed due to the combination of the P impurity surrounded the Titanium (4+) (Ti^{4+}) in the ATN crystal lattice. For 3D structure formation, the ATN-BP hybrid was

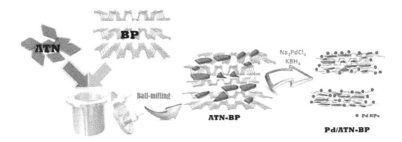

FIGURE 5.9   The formation Scheme of 3D Pd/ATN/BP (PdNPs anchored on anatase titanium dioxide-BP hybrids 3D structure). Adapted with permission from Wu et al. (2018), Copyright © 2017, WILEY–VCH Verlag GmbH & Co. KGaA, Weinheim.

used/introduced for the maintenance PdNPs. The resulting combination of Pd/ATN-BP-based 3D hybrids structure (with hetero borders of BP, Pd, and ATN) demonstrated and extraordinary enhanced the catalyst's electrolyte dispersion (durability for the EOR) and electron transport. Furthermore, through the durable synergistic interface among the ATN-BP and PdNPs, Pd/ATN-BP 3D hybrids structures also improved the taking off of the reactive portion of intermediates.

After completing the process of synthesis, the material TEM was used to monitor the material's internal morphology. Figure 5.10a–c shows the internal morphology of ATN-BP hybrids using the TEM images and found that the irregular small cataclastic ATN was formed as a result of the deformation of the rectangular ATN. The composition arrangement is confirmed from the brightness variance reflects and it is concluded that the catalectic ATN holds a position between the stable 3D ATN-BP hybrids material and BP nanoflakes (see inset of Figure 5.10d). Energy dispersive spectroscopy (EDS) and TEM images show the line-scan analysis of the 3D ATN-BP hybrids material having the EDS spectral confirm the presence of Ti and P, which are basic elemental compositions in the structure of the hybrid of ATN-BP (Figure 5.10e–g).

After analysis, it is very clearly detected that in the area of BP layers with small-size consistent catalectic ATN the P is less crowded than Ti. This evidence strongly suggested that there is a connection between the small-size cataclastic ATN and 3D hybrid ATN-BP-based structures with BP nanosheets. Furthermore the HRTEM image was used to confirm the presence of BP and ATN lattice fringes, the presence of d-spacing (of 0.335 and 0.262 nm, equivalent to the (021) and (040) interplanar space) confirms the presence of the BP nanoflakes and the lattice fringes (of 0.352 and 0.19 nm match to the (101) and (200) faces) relates to ATN, respectively (Figure 5.10h) (Wu et al. 2018).

5.3   COMPARISON OF 3D STRUCTURES OF BP WITH OTHER 3D STRUCTURES OF 2D LAYERED MATERIALS

5.3.1   3D STRUCTURES OF GRAPHENE

Different techniques have been developed in the past decade for improving the synthetic protocol of 3D graphene macrostructures (with a unique network of structure

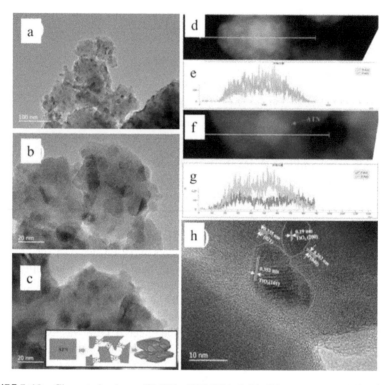

FIGURE 5.10 Characterizations of PdNPs-ATN-BP hybrids 3D structures. a–c) The internal morphology of ATN-BP hybrids using the TEM images; the irregular small cataclastic ATN was formed as a result of deformation of the rectangular ATN. The composition arrangement is confirmed from the brightness variance reflection. d) TEM image shows the internal morphology and cross-section of a separate ATN-BP hybrid structure material. e) Catalectic ATN hold a position between the non-reactive 3D constructions of the ATN-BP hybrids material and BP-based nanoflakes. f&g) The EDS-based line scan intersection (a piece of cataclastic consistent ATN) in an ATN-BP based hybrid structure material. h) HRTEM picture showing the confirmation of the development of an ATN-BP based hybrid material (inset of c). Adapted with permission from Wu et al. (2018), Copyright © 2017, WILEY–VCH Verlag GmbH & Co. KGaA, Weinheim.

in the form of large specific surface area, ultrahigh porosity, outstanding mechanical flexibility, and excellent electrical/thermal conductivities) based on the chemical vapor deposition and chemical exfoliation. Due to the above-mentioned characteristics, the potential applications and greatly extended functionalities of graphene are expected provide substantial benefits to the suggested 3D network based on graphene structures. The leading approaches for assembly and integration of the 3D network of graphene structures and shortcomings are summarized in this section. The main focus includes the template-based organization of the graphene sheets network, the organization and self-assembly of chemically resulting graphene-based nanosheets with the contribution of cross-linking like agents or during reduction, 3D printing, the gas expansion-based method of graphene films, template-engaged-based CVD development on different 3D porous structure-like substrates, and so on (Shehzad et al. 2016).

5.3.1.1 3D Graphene Network from GO-Derived Structures

During the past decade, different strategies were developed based on the CVD-grown graphene and chemical exfoliated for the synthesis process of 3D structures of graphene. For example, template-based assembly for the fabrication of graphene nanosheets, self-assembly as a result of chemically derived (during the reduction process) graphene sheets, the process of the template-based CVD development process on different substrates such as 3D porous metal/insulator, 3D printing, and gas expansion of graphene films etc. (Shehzad et al. 2016; Chen et al. 2019).

When we compared the above-mentioned strategies in terms of cast yield and simplicity, it was found that the synthetic strategies-based chemical exfoliated procedures of graphene are most encouraging due to low costs and scale-up manufacture of 3D structures of graphene during the processing. Additionally, for conversion into several 3D graphene-based macroscopic organizations using the self-assembly-based approach of GO sheets or template-directed assembly approach, the chemical exfoliated GO sheets have been largely used due to their easy processing approach, and 3D graphene porous structures are formed as a result of the gas expansion inside GO papers (Chen et al. 2019).

5.3.1.1.1 Solution-Based Self-Assembly of GO Sheets

GO sheets can be distributed in water due to the richness of hydrophilic oxygen-containing functional groups, and the electrostatic forces of repulsion between the oxygen functional groups producing a highly stable and concentrated aqueous solution. Meanwhile, the GO basal plane's highly hydrophobic plane is due to the 2D engagements of sp^2 carbon structures. The equilibrium among rGO platelets and GO due to the presence of van der Waals force representing their self-assembly performance in water containing liquid media and regulates their solution properties. Synthesis of porous macrostructures and 3D graphene hydrogel can be controlled through the organization of the GO colloidal and rGO (Chen et al. 2019).

5.3.1.1.1.1 Cross-Linking GO-Based Gelation   As a result of the process of gelation between the GO sheets and the cross-linking mediators—such as DNA (Xu, Wu, et al. 2010), protein (Huang et al. 2011), polymeric material (Bai et al. 2010; Worsley et al. 2010), small quaternary ammonium salts, noble metal nanocrystal (Tang et al. 2010), and metal ions—self-assemble structures of 3D graphene hydrogel were achieved (Figure 5.11a) (Bai et al. 2011; Cong et al. 2012; Jiang et al. 2010).

Bai et al. (2010) introduced polyvinyl alcohol (PVA) as a model cross-linking mediator and reported a graphene-based hydrogel. PVA and GO aqueous solutions were mixed together, and the resulting mixture was then aggressively stirred for around 10 seconds and sonicated for around 20 minutes to form PVA functionalized GO hybrid hydrogel.

There is a strong hydrogen bond between the resultant hydrogel for the cross-linking sites containing the oxygen-containing functionalities on the surface of

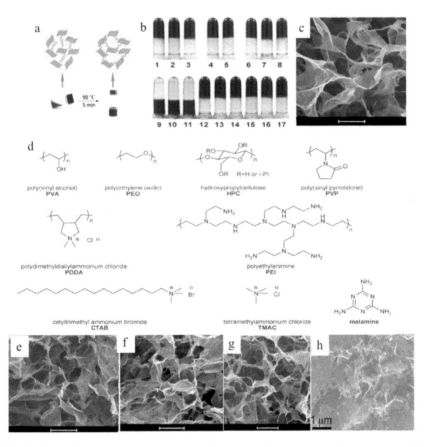

FIGURE 5.11 Characterization process of GO cross-linking process in solution. a) The proposed gelation mechanism and the process for synthesizing GO/DNA hydrogel. Adapted with permission from Xu, Wu, et al. (2010), Copyright © 2010, American Chemical Society. b) Snapshot of mixing of different cross-linkers with 5 mg/mL GO solutions: 1) (PVP) poly(vinylpyrrolidone) 0.5 mg/mL; 2) (HPC) hydroxypropyl cellulose 1 mg/mL; 3) (PEO) poly(ethylene oxide) 1 mg/mL; 4) (PDDA) polydimethyldiallylammonium chloride 0.1 mg/mL; 5) (PEI) polyethyleneimine 0.2 mg/mL; 6) (CTAB) cetyltrimethylammonium bromide 0.3 mg/mL; 7) (TMAC) tetramethylammonium chloride 1.9 mg/mL; 8) melamine 0.3 mg/mL; 9) Li^+ with a concentration value of 20 mM; 10) K^+ with a concentration of 20 mM; 11) Ag^+ with a concentration of 20 mM; 12) Mg^{2+} with a concentration of 15 mM; 13) Ca^{2+} with a concentration of 9 mM; 14) Cu^{2+} with a concentration of 3 mM; 15) Pb^{2+} with a concentration of 3 mM; 16) Cr^{3+} with a concentration of 3 mM; and 17) Fe^{3+} with a concentration of 3 mM. c) SEM pictures of GO solution. d) Structure of different cross-linker used for cross-linking in GO solution. e–h) Three characteristic GO hydrogels and lyophilized GO solution: e) 0.1 mg/mL with PDDA with GO/PDDA hydrogel; f) 1 mg/mL PVP combine with GO/PVP hydrogel; g) 9 mM Ca^{2+}. CGO ¼ 5 mg/mL, and scale bar length ¼ 10 mm g with GO/Ca^{2+} hydrogel (Bai et al. 2011). h) SEM image of sol-gel polymerized phenolic resin for the preparation of a graphene-based aerogel cross-linked. Adapted with permission from Worsley et al. (2010), Copyright © 2010, American Chemical Society.

GO sheets and the hydroxyl-rich PVA chains. Additionally, GO-based composite hydrogel production is possible due to the development of an unbroken GO-based network structure after increasing the number of cross-linking sites (Figure 5.11b). Different supramolecular interactions (such as p-stacking, electrostatic interaction, hydrogen bonding) and coordination is seen after introducing different cross-linking agent such as (polymers, ionic species, organic small molecules) (Figure 5.11d) in graphene oxide solution (Figure 5.11c) for further conducting the 3D organization of GO-based nanosheets in aqueous-solvent media in a systematic way to form hydrogels (Figure 5.11e–h) (Bai et al. 2011).

In another study, Worsley et al. (2010) synthesized a graphene-based hydrogel as a result of cross-linking between the organic sol-gel polymerization of phenolic resin and GO sheets (Figure 5.11h). Consequently, the resulted pyrolysis and supercritical dry of the hydrogel in an ultra-low-density graphene-based composite aerogel had a large value of surface areas (584 m^2/g) and high conductivity (1 S/cm) (Worsley et al. 2010).

5.3.1.1.1.2 Reduced GO-Based Self-Assembly For the preparation process of 3D macrostructures of graphene hydrogel, the technique of self-assembly based on the cross-linking agent method has been widely used and has shown great success. As a result of the cross-link process at the reduction stage, the porous macrostructures and the 3D pristine graphene hydrogel (additive-free) can be produced. A large number of oxygen-containing functional groups (hydrophilic in nature) on the surface of GO-based nanosheets were randomly distributed in water before reduction due to the strong and durable electrostatic based dislike consequences. Due to the removal of hydroxyl and epoxy groups after beginning the reduction of GO sheets, the p-conjugated constructions of the GO basal plane and the hydrophobicity characteristic both increase. As a result of increasing P–P stacking between graphene and van der Waals forces, there is an increase in the number of cross-links between the adjacent GO sheets and 3D cross-linked graphene networks. Xu et al. first time reported the reduction-induced gelation of GO (Xu, Sheng, et al. 2010) by introducing a hydrothermal reduction process (Figure 5.12a–c). The treatment process of GO aqueous-based distribution was closed in a Teflon-lined autoclave at a temperature of 180°C for time of 12 hours. Consequently, well-defined 3D porous structures of rGO hydrogel are formed. There is an extraordinary electrical conductivity value of 5103 S/cm and durable storage modulus of 450–490 kPa is found in the rGO-based hydrogel. For the preparation purpose of different types of 3D graphene doping and macroporous hybrid structures, the hydrothermal reduction method was developed (Zhao et al. 2012; Wu, Winter, et al. 2012; Zhao, Ren, and Cheng 2012). The approach demonstrated the hydrothermal management of pyrrole and GO varied suspension in a reaction vessel made of Teflon-lined autoclave at a temperature of 180°C for a time duration of 12 hours (Figure 5.12 d); as a consequence the ultralight hybrid (nitrogen (N)-doped graphene) network with a large value of surface area (280 m^2/g), a high value of electrical conductivity (12 S/cm), and density value of 2.1 mg/cm^3 is obtained (Zhao et al. 2012). On the contrary, in another study Bi et al. (2012) reported a pH-facilitated hydrothermal-based reduction route of GO dispersion to get the compacted high-density graphene macrostructures to have different controllable

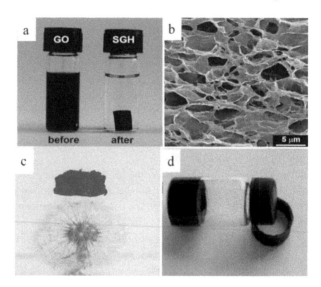

FIGURE 5.12 The reduction process of the self-assembly of graphene oxide (GO). a) Snapshot at 180°C before and after hydrothermal reduction process for a duration of 12 hours of a 2 mg/mL homogeneous GO aqueous dispersion. b) Surface morphology of the graphene hydrogel interior microstructures are analyzed by SEM image. Adapted with permission from Xu, Sheng, et al. (2010), Copyright © 2010, American Chemical Society. c) Snapshot of as-prepared graphene-based network. Adapted with permission from Zhao et al. (2012), Copyright © 2012, WILEY–VCH Verlag GmbH & Co. KGaA, Weinheim. d) The graphene hydrogel in a reaction flask is synthesized as a result of providing heat content to the mixture of L-ascorbic acid without any stirring, graphene oxide macrostructures form as a result. Adapted with permission from the Royal Society of Chemistry (Zhang, Sui, et al. 2011)

shapes. The solid microstructure of compacted graphene influenced with a high-temperature treatment under annealing at 900°C, high density (~1.6 g/cm³), further the electrical conductivity and compressive strength value of the compact graphene 3447.6 S/cm 361 MPa, respectively. The value of density was comparable to ultra-high compressive strength and conventional graphite products.

5.3.1.1.2 Gas Extension Technique

Inside the graphene film, the gas expansion creates space between the graphene sheets and some pores; as a result, 3D porous graphene structures are formed. As an example, El-Kady et al. (2012) introduced the direct reduction of GO films for the preparation of 3D porous graphene films (Figure 5.13) by a laser scribing method. For that purpose, a vacuum filtering or drop-casting of GO is introduced for the preparation of a GO-based thin layer film onto a flexible substrate such as PET or PDMS, and the GO-based thin layer film was firmly attached on a maximum of a Light Scribe-allowed DVD media. After the film's formation, the disc is rotated for laser treatment into the DVD optical drive. As a result of the oxidation process (oxygen functional groups removed as a gaseous species), the sp² (unsaturated) carbons are

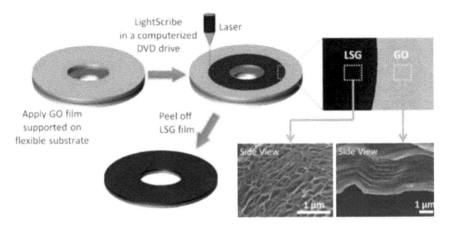

FIGURE 5.13 The fabrication of graphene (LSG) laser-scribed films as a result of the gas expansion technique on 3D porous graphene structures. On the highest Light Scribe-allowed DVD media disc, VA graphene oxide (GO) film sustained on the surface of a flexible substrate. The snapshot shows the GO film change in color as a result of reduction through the laser-scribed structure of graphene. The SEM pictures show the well-exfoliated few-layered porous graphene film. Adapted with permission from El-Kady et al. (2012), Copyright © 2012, American Association for the Advancement of Science.

formed after the laser irradiation process. As a result, there is a variation in the conductivity of the thin layer film (conductivity value is increased to 17.38 S/cm from the GO (insulating) to graphene (conducting nature). Through laser irradiation, there is a rapidly evolving process of the gaseous species, conversion into the porous well-exfoliated structure of graphene-based films with a precise maximum value of surface area 1520 m²/g from stacked GO films and variation in morphology is monitored by SEM images (Figure 5.13) (Chen et al. 2019).

5.3.2   3D STRUCTURES OF TRANSITION METAL DICHALCOGENIDES

5.3.2.1   MoS₂

Recent global crises regarding energy and pollution control mean that there is extraordinary demand for the design and construction of a large number of devices for electrical energy storage. Among the considered approaches, lithium-ion batteries (LIBs) act as a promising candidate for portable electronics due to their excellent security, long cycle life, and high energy density (Goodenough and Kim 2009; Choi, Chen, et al. 2012; Kang, Chen, and Ogunseitan 2013; Zhang et al. 2016). Tian et al. (2016) used a one-pot hydrothermal method for preparing 3D composites of layered MoS_2/interconnected graphene nanoribbons (GNRs). The GNRs were used as a conductive linker or bridge between MoS_2 layers to form the 3D structure using solution-processed self-assembly due to electrostatic attraction. The composites have plentiful mesopores and large surface areas, which provide a direction to the suitable contact place between the electrolyte and the electrode. To enhance the value of specific

FIGURE 5.14   3D MoS_2/PANI hierarchical nanoflower-like structures under annealing in an inert atmosphere at 500°C for four hours. a&b) SEM images show the surface morphology of the 3D structures of MoS_2/PANI hierarchical structures. d&e) Surface morphology of nanoflowers. c&f) Snapshot of two different kinds of Chinese roses. Adapted with permission from Hu et al. (2014), Copyright © 2014, American Chemical Society.

capacity, the Li-ion could simply add and eliminate the active materials. Additionally, for fast electron transportation process in Li-ion, the combined effects of the MoS_2 and the GNRs layers provide better results in accommodating the expansion of volume to gain the rated capacity, and cycle stability prevents the self-aggregation of MoS_2 nanosheets. Finally, the MoS_2/GNRs composite anode has excellent electrochemical properties. After measuring the quantity of 80 cycles by the cycling test machine and 606.8 mA hg^{-1} at 3 Ag^{-1} by the rate competence test, the treated composites have a specific capacity value of 1009.4 mA hg^{-1} at 200 mAg^{-1}.Comparatively, MoS_2 has a specific capacity of 139.8 mA hg^{-1} at 200 mAg^{-1}after a number of 80 cycles by the cycling test and 37.4 mA hg^{-1} at 3 Ag^{-1} by the rate capability test. Hu et al. (2014) presented a simple hydrothermal route to synthesize 3D PANI-functionalized MoS_2-based hierarchical nanoflower-like structures under annealing in an inert atmosphere at 500°C for four hours, as shown in Figure 5.14. The resultant composite MoS_2/PANI and MoS_2/C nanoflower-like structures are constructed of ultrathin structure-based nanoplates (consist of FL-MoS_2 nanosheets and PANI or carbon) confirmed by structural characterizations. Moreover, sufficient void space and the large specific area between the two adjacent nanoplates are provided by the 3D hierarchical and composite structures of PANI based MoS_2 (Figure 5.14a–b) and MoS_2/C (Figure 5.14d–e) nanoflowers, and both the structures are quite similar to a snapshot of two categories of Chinese roses (Figure 5.14c, f) (Hu et al. 2014). Additionally, sufficient void space and a large specific area between the adjacent nanoplates play a key role for the improvement of the electrochemical performance. Further, a 3D composite of MoS_2/ C hierarchical structures can be constructed through a simple and robust annealing process which acts as a supreme candidate for presentation in lithium-ion batteries (LIBs) (Hu et al. 2014).

5.3.2.2 WS$_2$

Due to the encouraging electrocatalyst for hydrogen evolution reaction (HER), the Tungsten disulfide (WS_2) has been the cause of much consideration for scientists during the past five years. A lot of research has been published related to two-dimensional planar structures of WS_2 electrode materials (Voiry et al. 2013; Zhang, Wang, et al. 2015; Lin et al. 2014; Zhang, Liu, et al. 2015; Yang et al. 2013; Cheng et al. 2014; Wu, Fang, et al. 2012; Lukowski et al. 2014); due to that planner structure, there is a low catalyst loading responsible for a limit of the HER efficiency.

Qi et al. (2017) developed a huge number of active places containing the 3D structure of the WS_2/graphene/Ni electrode structure with good stability and without extra polymer binder (as a free binder hydrogen evolution reaction (HER) electrode) as shown in Figure 5.15a. The different steps involved in the preparation of electrode are as follows. In the initial step the precursor solution was prepared as a result of dissolution of $(NH_4)_2WS_4$ (with a quantity of 1.2 g) using N, N-Dimethylformamide (DMF) as a solvent (20 mL), and after that the solution was treated for sonication for 30 minutes to get the homogenous phase. In the second stage, the prepared (NH4)2WS4 solution is used to prepare a dispersion of the 3D graphene/Ni foam and then the resultant product is dried out by heat (at 80°C) for 30 minutes. Consequently, to get the synthesized product in dry form, the synthesized products were put in the quartz tube with stable Ar/H$_2$ flux of 100/100 cm and finally annealed at a temperature of 350°C for 30 minutes, as shown in Figure 5.15b–c (Qi, Li, et al. 2015). The morphology of synthesized product is monitored by SEM image analysis. Different SEM pictures of 3D structures of Ni (Figure 5.15d), 3D hybrid structure of graphene/Ni (Figure 5.15e), and 3D hybrid structure of WS_2/graphene/Ni (Figure 5.15f) showed the surface morphology of the synthesized product.

The studies disclose that the synthesized hybrid material 3D-based structure of WS_2/graphene/Ni has a high current density of 119.1 mA cm^2 at 250 mV overpotential and a low overpotential of 87 mV at 10 mA cm^2, indicating the high HER activity. Furthermore, its good stability is indicated through its current density value, which decreased (from upper-value range 119.1 mA cm^2 to lower value range 110.1 mA cm^2) after 3000 cycles. The robust contact between WS_2 and high amount of loading capacity of WS_2 based catalysts, graphene/Ni backbones, and unique graphene-based 3D structure are responsible for the outstanding HER efficiency of a 3D hybrid structure of WS_2/graphene/Ni. It is expected that the 3D hybrid structure of WS_2/ graphene/Ni can also be used as an encouraging candidate for HER efficient catalyst as well as being functional in other electrochemical-based devices for energy storage applications. Additionally until now, like WS_2, the 3D structures of WSe_2 have rarely been reported (Qi et al. 2017).

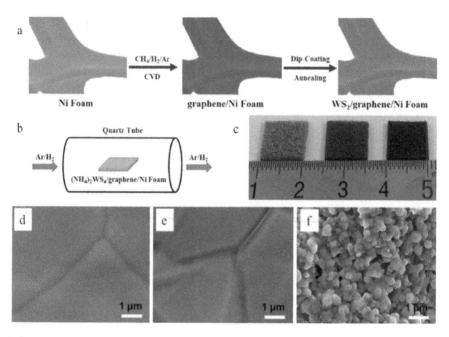

FIGURE 5.15 a) Different steps involved in preparing the 3D hybrid based structure of WS_2/graphene/Ni foam. b) Schematic representation and basic instrumentation of tube furnace used for the chemical synthesis process. c) Snapshot of different foam such as 3D Ni (left), 3D graphene/Ni (middle), and 3D hybrid structure of WS_2/graphene/Ni (right). SEM pictures showing the morphology of d) 3D Ni, e) 3D hybrid graphene/Ni, and f) 3D hybrid structure of WS_2/graphene/Ni. Adapted with permission from Qi et al. (2017), Copyright © 2021, all rights reserved.

5.3.3 3D STRUCTURES OF BORON NITRIDE

The main aim of 3D boron nitride (BN) nanomaterial is to regulate places/gaps in BN frameworks for the process of chemical conversion, adsorption, separation, etc. Generally, there are two commonly applied strategies—hard-templating and template-free strategies—that were used for obtaining the nonporous BN. Due to some shortcoming (due to complex boron chemistry), the process of soft templating based synthetic approaches are not fully explored for the BN system most widely adapted for nonporous carbons, despite some boron-incorporated block copolymers such as polynorbornene-decaborane utilized as a soft template and boron precursor. On the other hand, in case of hard-templating, by nano-casting on porous silica, there is a chance of obtaining a reverse porosity to the hard templates (Malenfant et al. 2007; Dibandjo, Chassagneux, et al. 2005; Rushton and Mokaya 2008; Alauzun et al. 2011; Bernard and Miele 2014; Lu et al. 2013), while a similar kind of porosity was accepted as a result of the elemental substitution of graphene aerogel (Figure 5.16a–f) (Rousseas et al. 2013) or porous carbons (Han et al. 2004; Vinu et al. 2005; Suryavanshi et al. 2014), and the caused definite surface area (SSA) value is fluctuating ranging from 100–950 m^2/g. Yang, Wang, et al. (2018) successfully suggest a

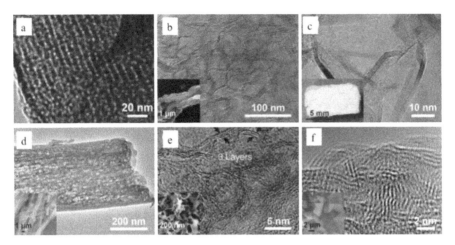

FIGURE 5.16 Nonporous boron nitride (BN) synthesized using a–c) hard-templating-based method and d–f) non-templating based technique: a) TEM pictures show the internal morphology structure of well-organized mesoporous BN-based template prepared from well-organized mesoporous carbon termed as CMK-3 using a nano-casting technique. Reprinted from Dibandjo, Bois, et al. (2005), Copyright © 2005, WILEY–VCH Verlag GmbH & Co. KGaA, Weinheim. b) The images of TEM (inset) of mesoporous BN template prepared from the mesoporous carbon using the elemental replacement technique. Reprinted from Vinu et al. (2005), Copyright © 2005, American Chemical Society. c) BN-based aerogel based template prepared from graphene-based structure of aerogel as a result of elemental substitution technique; the morphology is shown in TEM image (inset). Reprinted from Rousseas et al. (2013), Copyright © 2013, American Chemical Society. d) TEM and SEM (inset) pictures of the product obtained as a result of non-templating reaction of boric acid-melamine (2B M) originators of porous BN. Reprinted from Weng et al. (2013), Copyright © 2013, American Chemical Society. e) Scanning electron microscopy (SEM) and high resolution (HRTEM) (inset) pictures of porous BN-based nanosheets obtained as a result of a template-free-based reaction of B_2O_3 and guanidine hydrochloride. Reproduced under the terms of the Creative Commons Attribution 4.0 International. Copyright © the authors, published by Springer Nature (Lei et al. 2013). f) The result of a template-free reaction of dicyanamide and boric acid, the HRTEM images and SEM (inset) pictures of porous structures of BN microsponges. Reprinted from Weng et al. (2014), Copyright © 2013, WILEY–VCH Verlag GmbH & Co. KGaA, Weinheim.

new synthesis strategy for producing flower-like morphology-controlled 3D boron nitride nanostructures (BNNSs).

Guanidine hydrochloride (~3.95 g) and boron oxide (~0.56 g) were combined in 10 mL methanol and, to form a homogeneous solution, stirred for 24 hours continuously. After drying the originator mixture (~1.5 g) in the air, the mechanical force is used for the resultant crystalline powders and afterwards crushed to a cylinder having a radius value of ~5 mm and height 15 mm, after applying an amount of 3 tons of force for around five minutes. The mixture of precursor in the cylinder was preserved for heat up to 1050°C with a constant heating rate value of 1.5°C/min and continuously

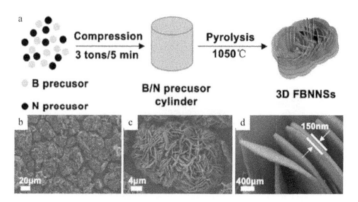

FIGURE 5.17 a) Schematic procedures of 3D functionalized structures of the flower-like shape of boron nitride nanosheets (FBNNSs) synthesis process. b) Low-resolution and c&d) high-resolution SEM pictures showing the surface morphology of BP nanosheets (Yang, Wang, et al. 2018).

supplying the same temperature for two hours under inert atmosphere after supplying the N_2 gas with controlled pebbling per unit time. During the synthesis process, the guanidine hydrochloride and boron oxide are used as corresponding N and B sources. They have compacted into N/B or B/N precursor cylinder (Figure 5.17a). The pyrolysis process is introduced for 3D functionalized FBNNSs via under the inert atmosphere and low heating rate in the second stage. SEM images show the unique 3D flower-like surface morphology (Figure 5.17b–c) and the resultant products in SEM showed that the size is approximately in ranging from 10 to 100 μm. Simultaneously, the value of thickness is found around ~150 nm as mentioned in Figure 5.17d.

Due to unique properties, a 3D structure such as hierarchical porosities and high pore volume, high specific surface area, as well as having terminated polar functional groups (–OH and –NH_2) the 3D functionalized FBNNSs is used as a strong candidate for NH_3 and CO_2 adsorption capacity value in association with original h-BN. Moreover, nonporous BN could be utilized in multidirectional applications, including such as drug delivery, for H_2 packing, pollutant management, catalyst maintenance, etc.(Yang, Wang, et al. 2018).

Nowadays, there are many challenges and there is much more to do to significantly improve BN nanomaterial quality, such as considering the scalable yield of syntheses and gaining a full understanding of their functional potential value within the field of BN nanomaterial.

5.3.4 Comparison of 3D Structures of BP with Other 3D Structures of 2D Layered Materials

BP has attracted much research interest recently, due to its distinctive 2D geometry with puckered geometry, the high value of carrier mobility, great value of the surface area, and strong capacity of mechanical strength, particularly in case of energy storage and transformation uses, its high efficiency of photothermal transformation (Shao et al. 2016; Sun et al. 2015), a high chemical reactivity that might be responsible for a high amount of drug-packing capability and as a result increased the capacity of loading (Chen, Ouyang, et al. 2017), and excellent biodegradation and biocompatibility (Shao et al. 2016). Recently, due to its semiconducting properties that graphene may not possess, single-layer phosphorene has stimulated much more research interest. Moreover, studies showed that few-layers of single-layer phosphorene possess high carrier mobility, optical and electronic anisotropy, and a tunable bandgap (Kou, Chen, and Smith 2015; Ling et al. 2015), which graphene does not possess, giving phosphorene a competitive edge and potentially stimulating new innovations in 2D materials. Furthermore, phosphorene has higher carrier mobility than TMDs and due to that high carrier mobility, it is expected to function as a great bridge between graphene and TMDs (Wen et al. 2019; Xu et al. 2013). The TMDs, like Tungsten disulfide (WS_2) (Kou, Chen, and Smith 2015) and molybdenum disulfide (MoS_2) (Liu, Du, et al. 2015; Wen et al. 2019), show the nonlinear value of optical response, which is stronger than the graphene (Xu et al. 2013; Naguib et al. 2014; Wang et al. 2013). Additionally, WS_2 has a large nonlinear refractive index (Zheng et al. 2015), a high threshold value of optical damage (Mao et al. 2015), and a large nonlinear second-order susceptibility (Janisch et al. 2014). The layer-d, dependent direct bandgap varying from bulk (0.35 eV) to monolayer (2 eV) makes BP accomplished in filling in the gap between wide bandgap TMDs and semi-metallic graphene (Qiao et al. 2014). Therefore, due to these multidirectional properties, BP can be used as an alternative material for minimizing the hurdles for designing ultrathin large-area membranes for channels in the silicon semiconductor industry for the preparation of 3D hybrid structures (Akahama, Endo, and Narita 1983; Doganov et al. 2015). The BP laminated layer/membrane-based 3D hybrid structures have the great van der Waals (vdWs) interaction for other 2D layered material, quantum dots, polymers, and small organic molecules, and act as good host/hybrid structures, with enhanced electrical, mechanical, biodegradable properties. In addition, improvement in synthetic protocol and assembly to 3D macrostructures of BP for technological potential applications is in development. It is analyzed that the 3D macroscopic structures of 2D materials are porous in nature, so can be used as electrodes for batteries/supercapacitor applications (Chen 2013; Wang et al. 2010; Guo, Song, and Chen 2009; Hou et al. 2011; Simon and Gogotsi 2008; Wang, Zhang, and Zhang 2012; Zhang and Zhao 2009). Moreover, the pores can host a variety of other materials including metals, organic molecules, polymers, etc. (Dong et al. 2012a, 2012b; Yong et al. 2012; Sattayasamitsathit et al. 2013; Maiyalagan et al. 2012) resulting in new functionalities and applications. 3D macrostructures of black phosphorus (BP) will not only contain the fundamental characteristics of BP such as high mobility, tunable narrow bandgap, and mechanical strength, but also contain new functionalities due to their porous structure. These pores provide the facilitation for ion transport and diffusion, and the interconnecting of 2D nanosheets in the formation process of 3D structure provides abundant paths for electron transport. These attributes are important for efficient energy conversion and storage materials. So, from the above discussion we can conclude that 2D nanosheets of BP materials may have application in electronic devices, but for energy conversion/storage and other biomedical application where porous large-scale structures are required, 3D macrostructures of BP can come in handy (Lv et al. 2018; Wen et al. 2019; Yang, Yin, et al. 2018; Xing et al. 2018). The details are outlined in Table 5.1.

TABLE 5.1
Detail summary of BP with other 2D material

No	Type of 2D material	Invention year	Composition and structure	Materials/bandgap (eV)	Carrier mobility $(cm^2V^{-1}s^{-1})$	On-off current Ratio	Electrical conduction type	Biocompatibility behavior	In vivo Biodegradability behavior	Major Methods of preparation	Major application	Reference
1	Black phosphorus (BP)	2014	Puckered structure and three crystalline shaped geometries: simple cubic, orthorhombic, and rhombohedral.	Semiconductor/ 0.3–2.0	1000	10^3–10^5	Ambipolar	Yes	Yes	Different top-down and bottom-up approaches were developed for the synthesis of BP based nanomaterial	Solar cells; Energy storage devices, photodetection devices	Ospina et al. 2016; Kou, Chen, and Smith 2015; Li et al. 2014
2	Graphene	2004	Carbon; hexagonal honeycomb structure	Semimetal/ 0	2×105	5.5–44	Ambipolar	Yes	No, need further functionalization for the preparation of biodegradable material	Exfoliation-based synthesis (liquid, mechanical, and chemical), template-based and template free approaches	Electronics and photonics devices, Gas and bio-sensors; Thermal, and Mechanical applications; Electrochemical Energy storage applications	(Choi et al. 2010; Khan et al. 2015; Jiao et al. 2009; Han 2009; Akinwande, Petrone, and Hone 2014; Das et al. 2014)
3	MoS_2	2010	Molybdenum-Sulphur TMDCs; a hexagonal structure having Mo and S_2 atoms positioned at irregular corners	Semiconductor/ 1.2–1.8	10–200	106–108	n-type	Yes	No, need further functionalization for the preparation of biodegradable material	Exfoliation-based synthesis (liquid, mechanical, and chemical), template-based and template free approaches	Supercapacitor; Photodetector; memory device Photovoltaic device, chemical sensing	(Li and Zhu 2015; Kou, Chen, and Smith 2015; Wang et al. 2012b; Sahoo et al. 2013)

#	Name	Year	Structure	Type/Band gap					Functionalization	Synthesis	Applications	References
4	WS$_2$	2012	Tungsten, Sulphur WS2 having the geometry like planar structures	Semiconductor/ 1.3–2.1	1000	10^5	(n-type)	Yes	No, need further functionalization for the preparation of biodegradable material	Exfoliation-based synthesis (liquid, mechanical, and chemical), template-based and template free approaches	photocatalysts and optoelectronics such as field-effect transistors and solar cells	(Elías et al. 2013; Lan et al. 2015; Wang et al. 2012a; Wang et al. 2012b; Huo et al. 2015)
5	WSe$_2$	2012	Tungsten, selenium (WSe$_2$); Hexagonal geometry like structure due to location of W and Se$_2$ atoms at alternating corners	Semiconductor/ 1.2–1.7	140–500	10^4–10^6	Ambipolar	Yes	No, need further functionalization for the preparation of biodegradeable material	Mechanical exfoliation process, chemical vapour deposition process	Photovoltaic:Photoelectrochemical (PEC) devices, FET	(Liu, Fathi, et al. 2015; Fang et al. 2012; Yu et al. 2015; Akinwande, Petrone, and Hone 2014)
6	Nanosheets based on hexagonal boron nitride (h-BNNS)	2004	Boron nitride having the hexagonal honeycomb-like structure	Insulator/ ~5.9	-	-	Ambi-polar	Yes	No, need further functionalization for the preparation of biodegradable material	Exfoliation-based synthesis (liquid, mechanical and chemical), template-based and template free approaches	Nanoelectronics-ultrathin insulating films, catalysis	(Nag et al. 2010; Pacile et al. 2008; Coleman et al. 2011; Khan et al. 2015; Dean et al. 2012; Han 2009; Jo et al. 2013; Song et al. 2010)

5.4 SUMMARY

To summarize, 3D heterostructures are expected to be extensively studied in the coming years in order to develop effective synthetic methods and gain a better understanding of their properties and applications. During recent years, developments in BP-based 3D heterostructure synthesis strategies have been extremely promising, but they are still in their infancy. Also coordinated is the comparison of BP to other 3D layered material structures and BP is defined for potential and biomedical applications. Future research could resolve the lack of large-area standardized self-assembly and scalable routes for system fabrication on versatile substrates.

5.5 ACKNOWLEDGMENTS

The research was partially supported by financial support from the Science and Technology Development Fund (Nos. 007/2017/A1 and 132/2017/A3), Macao Special Administration Region (SAR), China, and National Natural Science Fund (Grant Nos. 61875138, 61435010, 61805147, and 6181101252), and Science and Technology Innovation Commission of Shenzhen (KQTD201503241627 0385, JCYJ20150625103619275, JCYJ20180305125141661, and JCYJ2017081 1093453105). The authors also acknowledge the support from the Instrumental Analysis Center of Shenzhen University (Xili Campus).

REFERENCES

Akahama, Yuichi, Shoichi Endo, and Shin-ichiro Narita. 1983. "Electrical properties of black phosphorus single crystals." *Journal of the Physical Society of Japan* 52(6): 2148–2155.

Akinwande, Deji, Nicholas Petrone, and James Hone. 2014. "Two-dimensional flexible nanoelectronics." *Nature Communications* 5: 5678.

Alauzun, Johan G., Simona Ungureanu, Nicolas Brun, Samuel Bernard, Philippe Miele, Rénal Backov, and Clément Sanchez. 2011. "Novel monolith-type boron nitride hierarchical foams obtained through integrative chemistry." *Journal of Materials Chemistry* 21(36): 14025–14030.

Bai, Hua, Chun Li, Xiaolin Wang, and Gaoquan Shi. 2010. "A pH-sensitive graphene oxide composite hydrogel." *Chemical Communications* 46(14): 2376–2378.

Bai, Hua, Chun Li, Xiaolin Wang, and Gaoquan Shi. 2011. "On the gelation of graphene oxide." *The Journal of Physical Chemistry C* 115(13): 5545–5551.

Baumann, Bernhard, Tomasz Jungst, Simone Stichler, Susanne Feineis, Oliver Wiltschka, Matthias Kuhlmann, Mika Lindén, and Jürgen Groll. 2017. "Control of nanoparticle release kinetics from 3D printed hydrogel scaffolds." *Angewandte Chemie International Edition* 56(16): 4623–4628.

Bernard, S, and Philippe Miele. 2014. "Nanostructured and architectured boron nitride from boron, nitrogen and hydrogen-containing molecular and polymeric precursors." *Materials Today* 17(9): 443–450.

Bi, H., K. Yin, X. Xie, Y. Zhou, N. Wan, F. Xu, F. Banhart, L. Sun, and R. S. Ruoff. 2012. Advanced Materials, 24: 5124–5129.

Broussard, Joshua A., Nicole L. Diggins, Stephen Hummel, Walter Georgescu, Vito Quaranta, and Donna J. Webb. 2015. "Automated analysis of cell-matrix adhesions in 2D and 3D environments." *Scientific Reports* 5: 8124.

Castellanos-Gomez, Andres. 2016. "Why all the fuss about 2D semiconductors?" *Nature Photonics* 10(4): 202.

Chang, Chunyu, and Lina Zhang. 2011. "Cellulose-based hydrogels: present status and application prospects." *Carbohydrate Polymers* 84(1): 40–53.

Chen, Jiajun. 2013. "Recent progress in advanced materials for lithium ion batteries." *Materials* 6(1): 156–183.

Chen, Tao, Yuhua Xue, Ajit K Roy, and Liming Dai. 2013. "Transparent and stretchable high-performance supercapacitors based on wrinkled graphene electrodes." *ACS Nano* 8(1): 1039–1046.

Chen, Wansong, Jiang Ouyang, Hong Liu, Min Chen, Ke Zeng, Jianping Sheng, Zhenjun Liu, Yajing Han, Liqiang Wang, and Juan Li. 2017. "Black phosphorus nanosheet-based drug delivery system for synergistic photodynamic/photothermal/chemotherapy of cancer." *Advanced Materials* 29(5): 1603864.

Chen, Yuanxun, Kefeng Cai, Congcong Liu, Haijun Song, and Xiaowei Yang. 2017. "High-performance and breathable polypyrrole coated air-laid paper for flexible all-solid-state supercapacitors." *Advanced Energy Materials* 7(21): 1701247.

Chen, Zongping, Lei Jin, Wenjun Hao, Wencai Ren, and Hui-Ming Cheng. 2019. "Synthesis and applications of three-dimensional graphene network structures." *Materials Today Nano* 5, 100027.

Cheng, Liang, Wenjing Huang, Qiufang Gong, Changhai Liu, Zhuang Liu, Yanguang Li, and Hongjie Dai. 2014. "Ultrathin WS2 nanoflakes as a high-performance electrocatalyst for the hydrogen evolution reaction." *Angewandte Chemie International Edition* 53(30): 7860–7863.

Choi, Bong Gill, MinHo Yang, Won Hi Hong, Jang Wook Choi, and Yun Suk Huh. 2012. "3D macroporous graphene frameworks for supercapacitors with high energy and power densities." *ACS Nano* 6(5): 4020–4028.

Choi, Nam-Soon, Zonghai Chen, Stefan A. Freunberger, Xiulei Ji, Yang-Kook Sun, Khalil Amine, Gleb Yushin, Linda F. Nazar, Jaephil Cho, and Peter G. Bruce. 2012. "Challenges facing lithium batteries and electrical double-layer capacitors." *Angewandte Chemie International Edition* 51(40): 9994–10024.

Choi, Wonbong, Indranil Lahiri, Raghunandan Seelaboyina, and Yong Soo Kang. 2010. "Synthesis of graphene and its applications: a review." *Critical Reviews in Solid State and Materials Sciences* 35(1): 52–71.

Coleman, Jonathan N., Mustafa Lotya, Arlene O'Neill, Shane D. Bergin, Paul J. King, Umar Khan, Karen Young, Alexandre Gaucher, Sukanta De, and Ronan J. Smith. 2011. "Two-dimensional nanosheets produced by liquid exfoliation of layered materials." *Science* 331(6017): 568–571.

Cong, Huai-Ping, Xiao-Chen Ren, Ping Wang, and Shu-Hong Yu. 2012. "Macroscopic multifunctional graphene-based hydrogels and aerogels by a metal ion induced self-assembly process." *ACS Nano* 6(3): 2693–2703.

Das, Saptarshi, Wei Zhang, Marcel Demarteau, Axel Hoffmann, Madan Dubey, and Andreas Roelofs. 2014. "Tunable transport gap in phosphorene." *Nano Letters* 14(10): 5733–5739.

Dean, C., A. F. Young, L. Wang, I. Meric, G. H. Lee, K. Watanabe, T. Taniguchi, K. Shepard, P. Kim, and J. Hone. 2012. "Graphene based heterostructures." *Solid State Communications* 152(15): 1275–1282.

Dibandjo, P., F. Chassagneux, L. Bois, C. Sigala, and P. Miele. 2005. "Comparison between SBA-15 silica and CMK-3 carbon nanocasting for mesoporous boron nitride synthesis." *Journal of Materials Chemistry* 15(19): 1917–1923.

Dibandjo, Philippe, Laurence Bois, Fernand Chassagneux, David Cornu, J.-M. Letoffe, Bérangère Toury, Florence Babonneau, and Philippe Miele. 2005. "Synthesis of boron nitride with ordered mesostructure." *Advanced Materials* 17(5): 571–574.

Doganov, Rostislav A., Eoin C. T. O'Farrell, Steven P. Koenig, Yuting Yeo, Angelo Ziletti, Alexandra Carvalho, David K. Campbell, David F. Coker, Kenji Watanabe, Takashi Taniguchi, Antonio H. Castro Neto, and Barbaros Özyilmaz. 2015. "Transport properties of pristine few-layer black phosphorus by van der Waals passivation in an inert atmosphere." *Nature Communications* 6: 6647.

Dong, Xiaochen, Jingxia Wang, Jing Wang, Mary B. Chan-Park, Xingao Li, Lianhui Wang, Wei Huang, and Peng Chen. 2012. "Supercapacitor electrode based on three-dimensional graphene-polyaniline hybrid." *Materials Chemistry and Physics* 134(2–3): 576–580.

El-Kady, Maher F, Veronica Strong, Sergey Dubin, and Richard B Kaner. 2012. "Laser scribing of high-performance and flexible graphene-based electrochemical capacitors." *Science* 335(6074): 1326–1330.

Elías, Ana Laura, Néstor Perea-López, Andrés Castro-Beltrán, Ayse Berkdemir, Ruitao Lv, Simin Feng, Aaron D. Long, Takuya Hayashi, Yoong Ahm Kim, and Morinobu Endo. 2013. "Controlled synthesis and transfer of large-area WS2 sheets: from single layer to few layers." *ACS Nano* 7(6): 5235–5242.

Fang, Hui, Steven Chuang, Ting Chia Chang, Kuniharu Takei, Toshitake Takahashi, and Ali Javey. 2012. "High-performance single layered WSe2 p-FETs with chemically doped contacts." *Nano Letters* 12(7): 3788–3792.

Fraley, Stephanie I., Yunfeng Feng, Ranjini Krishnamurthy, Dong-Hwee Kim, Alfredo Celedon, Gregory D. Longmore, and Denis Wirtz. 2010. "A distinctive role for focal adhesion proteins in three-dimensional cell motility." *Nature Cell Biology* 12(6): 598.

Goodenough, John B., and Youngsik Kim. 2009. "Challenges for rechargeable Li batteries." *Chemistry of Materials* 22(3): 587–603.

Guo, Peng, Huaihe Song, and Xiaohong Chen. 2009. "Electrochemical performance of graphene nanosheets as anode material for lithium-ion batteries." *Electrochemistry Communications* 11(6): 1320–1324.

Guo, Zhinan, Han Zhang, Shunbin Lu, Zhiteng Wang, Siying Tang, Jundong Shao, Zhengbo Sun, Hanhan Xie, Huaiyu Wang, Xue-Feng Yu, and Paul K. Chu. 2015. "From black phosphorus to phosphorene: basic solvent exfoliation, evolution of Raman scattering, and applications to ultrafast photonics." *Advanced Functional Materials* 25(45): 6996–7002.

Han, Wei-Qiang. 2009. "Anisotropic hexagonal boron nitride nanomaterials: synthesis and applications." *Nanotechnologies for the Life Sciences*: Online https://doi.org/10.1002/9781119951438.eibc2643.

Han, Wei-Qiang, Richard Brutchey, T. Don Tilley, and A Zettl. 2004. "Activated boron nitride derived from activated carbon." *Nano Letters* 4(1): 173–176.

Hou, Junbo, Yuyan Shao, Michael W. Ellis, Robert B. Moore, and Baolian Yi. 2011. "Graphene-based electrochemical energy conversion and storage: fuel cells, supercapacitors and lithium ion batteries." *Physical Chemistry Chemical Physics* 13(34): 15384–15402.

Hu, Liangbing, Mauro Pasta, Fabio La Mantia, LiFeng Cui, Sangmoo Jeong, Heather Dawn Deshazer, Jang Wook Choi, Seung Min Han, and Yi Cui. 2010. "Stretchable, porous, and conductive energy textiles." *Nano Letters* 10(2): 708–714.

Hu, Lianren, Yumei Ren, Hongxia Yang, and Qun Xu. 2014. "Fabrication of 3D hierarchical MoS2/polyaniline and MoS2/C architectures for lithium-ion battery applications." *ACS Applied Materials & Interfaces* 6(16): 14644–14652.

Huang, Cancan, Hua Bai, Chun Li, and Gaoquan Shi. 2011. "A graphene oxide/hemoglobin composite hydrogel for enzymatic catalysis in organic solvents." *Chemical Communications* 47(17): 4962–4964.

Hultgren, Ralph, N. S. Gingrich, and B. E. Warren. 1935. "The atomic distribution in red and black phosphorus and the crystal structure of black phosphorus." *The Journal of Chemical Physics* 3(6): 351–355.

Huo, Nengjie, Zhongming Wei, Xiuqing Meng, Joongoo Kang, Fengmin Wu, Shu-Shen Li, Su-Huai Wei, and Jingbo Li. 2015. "Interlayer coupling and optoelectronic properties of ultrathin two-dimensional heterostructures based on graphene, MoS2 and WS2." *Journal of Materials Chemistry C* 3(21): 5467–5473.

Janisch, Corey, Yuanxi Wang, Ding Ma, Nikhil Mehta, Ana Laura Elías, Néstor Perea-López, Mauricio Terrones, Vincent Crespi, and Zhiwen Liu. 2014. "Extraordinary second harmonic generation in tungsten disulfide monolayers." *Scientific Reports* 4: 5530.

Jiang, Xu, Yanwen Ma, Juanjuan Li, Quli Fan, and Wei Huang. 2010. "Self-assembly of reduced graphene oxide into three-dimensional architecture by divalent ion linkage." *The Journal of Physical Chemistry C* 114(51): 22462–22465.

Jiao, Liying, Li Zhang, Xinran Wang, Georgi Diankov, and Hongjie Dai. 2009. "Narrow graphene nanoribbons from carbon nanotubes." *Nature* 458(7240): 877.

Jo, Insun, Michael Thompson Pettes, Jaehyun Kim, Kenji Watanabe, Takashi Taniguchi, Zhen Yao, and Li Shi. 2013. "Thermal conductivity and phonon transport in suspended few-layer hexagonal boron nitride." *Nano Letters* 13(2): 550–554.

Kang, Daniel Hsing Po, Mengjun Chen, and Oladele A. Ogunseitan. 2013. "Potential environmental and human health impacts of rechargeable lithium batteries in electronic waste." *Environmental Science & Technology* 47(10): 5495–5503.

Khan, Majharul Haque, Zhenguo Huang, Feng Xiao, Gilberto Casillas, Zhixin Chen, Paul J Molino, and Hua Kun Liu. 2015. "Synthesis of large and few atomic layers of hexagonal boron nitride on melted copper." *Scientific Reports* 5: 7743.

Kim, Daeil, Gunchul Shin, Yu Jin Kang, Woong Kim, and Jeong Sook Ha. 2013. "Fabrication of a stretchable solid-state micro-supercapacitor array." *ACS Nano* 7(9): 7975–7982.

Kou, Liangzhi, Changfeng Chen, and Sean C. Smith. 2015. "Phosphorene: fabrication, properties, and applications." *The Journal of Physical Chemistry Letters* 6(14): 2794–2805.

Lan, Changyong, Chun Li, Yi Yin, and Yong Liu. 2015. "Large-area synthesis of monolayer WS2 and its ambient-sensitive photo-detecting performance." *Nanoscale* 7(14): 5974–5980.

Lei, Weiwei, David Portehault, Dan Liu, Si Qin, and Ying Chen. 2013. "Porous boron nitride nanosheets for effective water cleaning." *Nature Communications* 4: 1777.

Li, Likai, Yijun Yu, Guo Jun Ye, Qingqin Ge, Xuedong Ou, Hua Wu, Donglai Feng, Xian Hui Chen, and Yuanbo Zhang. 2014. "Black phosphorus field-effect transistors." *Nature Nanotechnology* 9: 372–377.

Li, Meng, Xinjian Yang, Jinsong Ren, Konggang Qu, and Xiaogang Qu. 2012. "Using graphene oxide high near-infrared absorbance for photothermal treatment of Alzheimer's disease." *Advanced Materials* 24(13): 1722–1728.

Li, Xiao, and Hongwei Zhu. 2015. "Two-dimensional MoS2: properties, preparation, and applications." *Journal of Materiomics* 1(1): 33–44.

Li, Yulin, Yin Xiao, and Changsheng Liu. 2017. "The horizon of materiobiology: a perspective on material-guided cell behaviors and tissue engineering." *Chemical Reviews* 117(5): 4376–4421.

Lin, Jian, Zhiwei Peng, Gunuk Wang, Dante Zakhidov, Eduardo Larios, Miguel Jose Yacaman, and James M. Tour. 2014. "Enhanced electrocatalysis for hydrogen evolution reactions from WS2 nanoribbons." *Advanced Energy Materials* 4(10): 1301875.

Ling, Xi, Han Wang, Shengxi Huang, Fengnian Xia, and Mildred S. Dresselhaus. 2015. "The renaissance of black phosphorus." *Proceedings of the National Academy of Sciences* 112(15): 4523–4530.

Liu, Bilu, Mohammad Fathi, Liang Chen, Ahmad Abbas, Yuqiang Ma, and Chongwu Zhou. 2015. "Chemical vapor deposition growth of monolayer WSe2 with tunable device characteristics and growth mechanism study." *ACS Nano* 9(6): 6119–6127.

Liu, Chen-Wei, Feng Xiong, Hui-Zhen Jia, Xu-Li Wang, Han Cheng, Yong-Hua Sun, Xian-Zheng Zhang, Ren-Xi Zhuo, and Jun Feng. 2013. "Graphene-based anticancer nanosystem and its biosafety evaluation using a zebrafish model." *Biomacromolecules* 14(2): 358–366.

Liu, Han, Yuchen Du, Yexin Deng, and Peide D. Ye. 2015. "Semiconducting black phosphorus: synthesis, transport properties and electronic applications." *Chemical Society Reviews* 44(9): 2732–2743.

Lu, Jiong, Kai Zhang, Xin Feng Liu, Han Zhang, Tze Chien Sum, Antonio H. Castro Neto, and Kian Ping Loh. 2013. "Order–disorder transition in a two-dimensional boron–carbon–nitride alloy." *Nature Communications* 4: 2681.

Lukowski, Mark A., Andrew S. Daniel, Caroline R. English, Fei Meng, Audrey Forticaux, Robert J. Hamers, and Song Jin. 2014. "Highly active hydrogen evolution catalysis from metallic WS 2 nanosheets." *Energy & Environmental Science* 7(8): 2608–2613.

Luo, Shaojuan, Jinlai Zhao, Jifei Zou, Zhiliang He, Changwen Xu, Fuwei Liu, Yang Huang, Lei Dong, Lei Wang, and Han Zhang. 2018. "Self-standing polypyrrole/black phosphorus laminated film: promising electrode for flexible supercapacitor with enhanced capacitance and cycling stability." *ACS Applied Materials & Interfaces* 10(4): 3538–3548.

Lv, Zhisheng, Yuxin Tang, Zhiqiang Zhu, Jiaqi Wei, Wenlong Li, Huarong Xia, Ying Jiang, Zhiyuan Liu, Yifei Luo, and Xiang Ge. 2018. "Honeycomb-lantern-inspired 3D stretchable supercapacitors with enhanced specific areal capacitance." *Advanced Materials* 30(50): 1805468.

Maiyalagan, T., X. C. Dong, P. Chen, and X. Wang. 2012. "Electrodeposited Pt on three-dimensional interconnected graphene as a free-standing electrode for fuel cell application." *Journal of Materials Chemistry* 22(12): 5286–5290.

Malenfant, Patrick R. L., Julin Wan, Seth T. Taylor, and Mohan Manoharan. 2007. "Self-assembly of an organic–inorganic block copolymer for nano-ordered ceramics." *Nature Nanotechnology* 2(1): 43.

Mao, Dong, Yadong Wang, Chaojie Ma, Lei Han, Biqiang Jiang, Xuetao Gan, Shijia Hua, Wending Zhang, Ting Mei, and Jianlin Zhao. 2015. "WS 2 mode-locked ultrafast fiber laser." *Scientific Reports* 5: 7965.

Murphy, Sean V., and Anthony Atala. 2014. "3D bioprinting of tissues and organs." *Nature Biotechnology* 32(8): 773.

Nag, Angshuman, Kalyan Raidongia, Kailash P. S. S. Hembram, Ranjan Datta, Umesh V. Waghmare, and C. N. R. Rao. 2010. "Graphene analogues of BN: novel synthesis and properties." *ACS Nano* 4(3): 1539–1544.

Naguib, Michael, Vadym N. Mochalin, Michel W. Barsoum, and Yury Gogotsi. 2014. "Two-dimensional materials: 25th anniversary article: MXenes: a new family of two-dimensional materials." *Advanced Materials* 26(7): 992–1005.

Niu, Zhiqiang, Haibo Dong, Bowen Zhu, Jinzhu Li, Huey Hoon Hng, Weiya Zhou, Xiaodong Chen, and Sishen Xie. 2013. "Highly stretchable, integrated supercapacitors based on single-walled carbon nanotube films with continuous reticulate architecture." *Advanced Materials* 25(7): 1058–1064.

Ospina, D. A., C. A. Duque, J. D. Correa, and Eric Suárez Morell. 2016. "Twisted bilayer blue phosphorene: A direct band gap semiconductor." *Superlattices and Microstructures* 97: 562–568.

Pacile, D., J. C. Meyer, Ç. Ö. Girit, and A. Zettl. 2008. "The two-dimensional phase of boron nitride: Few-atomic-layer sheets and suspended membranes." *Applied Physics Letters* 92(13): 133107.

Pawlyn, Michael. 2013. "Push the limits of 3D printing." *Nature* 494(7436): 174.

Qi, Dianpeng, Zhiyuan Liu, Yan Liu, Wan Ru Leow, Bowen Zhu, Hui Yang, Jiancan Yu, Wei Wang, Hua Wang, and Shengyan Yin. 2015. "Suspended wavy graphene microribbons for highly stretchable microsupercapacitors." *Advanced Materials* 27(37): 5559–5566.

Qi, Fei, Pingjian Li, Yuanfu Chen, Binjie Zheng, Xingzhao Liu, Feifei Lan, Zhanping Lai, Yongkuan Xu, Jingbo Liu, and Jinhao Zhou. 2015. "Effect of hydrogen on the growth of MoS2 thin layers by thermal decomposition method." *Vacuum* 119: 204–208.

Qi, Fei, Pingjian Li, Yuanfu Chen, Binjie Zheng, Jingbo Liu, Jinhao Zhou, Jiarui He, Xin Hao, and Wanli Zhang. 2017. "Three-dimensional structure of WS2/graphene/Ni as a binder-free electrocatalytic electrode for highly effective and stable hydrogen evolution reaction." *International Journal of Hydrogen Energy* 42(12): 7811–7819.

Qiao, Jingsi, Xianghua Kong, Zhi-Xin Hu, Feng Yang, and Wei Ji. 2014. "High-mobility transport anisotropy and linear dichroism in few-layer black phosphorus." *Nature Communications* 5: 4475.

Rogers, John A, Takao Someya, and Yonggang Huang. 2010. "Materials and mechanics for stretchable electronics." *Science* 327(5973): 1603–1607.

Rousseas, Michael, Anna P. Goldstein, William Mickelson, Marcus A. Worsley, Leta Woo, and Alex Zettl. 2013. "Synthesis of highly crystalline sp²-bonded boron nitride aerogels." *ACS Nano* 7(10): 8540–8546.

Rushton, Ben, and Robert Mokaya. 2008. "Mesoporous boron nitride and boron-nitride-carbon materials from mesoporous silica templates." *Journal of Materials Chemistry* 18(2): 235–241.

Sahoo, Satyaprakash, Anand P. S. Gaur, Majid Ahmadi, Maxime J.-F. Guinel, and Ram S. Katiyar. 2013. "Temperature-dependent Raman studies and thermal conductivity of few-layer MoS2." *The Journal of Physical Chemistry C* 117(17): 9042–9047.

Sattayasamitsathit, S., Y. E. Gu, K. Kaufmann, W. Z. Jia, X. Y. Xiao, M. Rodriguez, S. Minteer, J. Cha, D. B. Burckel, C. M. Wang, R. Polsky, and J. Wang. 2013. "Highly ordered multilayered 3D graphene decorated with metal nanoparticles." *Journal of Materials Chemistry A* 1(5): 1639–1645.

Shao, Jundong, Hanhan Xie, Hao Huang, Zhibin Li, Zhengbo Sun, Yanhua Xu, Quanlan Xiao, Xue-Feng Yu, Yuetao Zhao, Han Zhang, Huaiyu Wang, and Paul K. Chu. 2016. "Biodegradable black phosphorus-based nanospheres for in vivo photothermal cancer therapy." *Nature Communications* 7: 12967.

Shehzad, Khurram, Yang Xu, Chao Gao, and Xiangfeng Duan. 2016. "Three-dimensional macro-structures of two-dimensional nanomaterials." *Chemical Society Reviews* 45(20): 5541–5588.

Shifu, Chen, Chen Lei, Gao Shen, and Cao Gengyu. 2005. "The preparation of nitrogen-doped photocatalyst TiO2– xNx by ball milling." *Chemical Physics Letters* 413(4–6): 404–409.

Simon, Patrice, and Yury Gogotsi. 2008. "Materials for electrochemical capacitors." *Nature Materials* 7(11): 845–854.

Song, Li, Lijie Ci, Hao Lu, Pavel B. Sorokin, Chuanhong Jin, Jie Ni, Alexander G. Kvashnin, Dmitry G. Kvashnin, Jun Lou, and Boris I. Yakobson. 2010. "Large scale growth and characterization of atomic hexagonal boron nitride layers." *Nano Letters* 10(8): 3209–3215.

Sun, Caixia, Ling Wen, Jianfeng Zeng, Yong Wang, Qiao Sun, Lijuan Deng, Chongjun Zhao, and Zhen Li. 2016. "One-pot solventless preparation of PEGylated black phosphorus nanoparticles for photoacoustic imaging and photothermal therapy of cancer." *Biomaterials* 91: 81–89.

Sun, Zhengbo, Hanhan Xie, Siying Tang, Xue-Feng Yu, Zhinan Guo, Jundong Shao, Han Zhang, Hao Huang, Huaiyu Wang, and Paul K. Chu. 2015. "Ultrasmall black phosphorus quantum dots: synthesis and use as photothermal agents." *Angewandte Chemie International Edition* 54(39): 11526–11530.

Suryavanshi, Ulka, Veerappan V. Balasubramanian, Kripal S. Lakhi, Gurudas P. Mane, Katsuhiko Ariga, Jin-Ho Choy, Dae-Hwan Park, Abdullah M. Al-Enizi, and Ajayan Vinu. 2014. "Mesoporous BN and BCN nanocages with high surface area and spherical morphology." *Physical Chemistry Chemical Physics* 16(43): 23554–23557.

Tang, Zhihong, Shuling Shen, Jing Zhuang, and Xun Wang. 2010. "Noble-metal-promoted three-dimensional macroassembly of single-layered graphene oxide." *Angewandte Chemie International Edition* 49(27): 4603–4607.

Tian, Ran, Weiqiang Wang, Yaolin Huang, Huanan Duan, Yiping Guo, Hongmei Kang, Hua Li, and Hezhou Liu. 2016. "3D composites of layered MoS2 and graphene nanoribbons for high performance lithium-ion battery anodes." *Journal of Materials Chemistry* 4(34): 13148–13154.

Vinu, Ajayan, Mauricio Terrones, Dmitri Golberg, Shunichi Hishita, Katsuhiko Ariga, and Toshiyuki Mori. 2005. "Synthesis of mesoporous BN and BCN exhibiting large surface areas via templating methods." *Chemistry of Materials* 17(24): 5887–5890.

Voiry, Damien, Hisato Yamaguchi, Junwen Li, Rafael Silva, Diego C. B. Alves, Takeshi Fujita, Mingwei Chen, Tewodros Asefa, Vivek B. Shenoy, and Goki Eda. 2013. "Enhanced catalytic activity in strained chemically exfoliated WS 2 nanosheets for hydrogen evolution." *Nature Materials* 12(9): 850.

Wang, Guoping, Lei Zhang, and Jiujun Zhang. 2012. "A review of electrode materials for electrochemical supercapacitors." *Chemical Society Reviews* 41(2): 797–828.

Wang, Hailiang, Li-Feng Cui, Yuan Yang, Hernan Sanchez Casalongue, Joshua Tucker Robinson, Yongye Liang, Yi Cui, and Hongjie Dai. 2010. "Mn3O4–graphene hybrid as a high-capacity anode material for lithium ion batteries." *Journal of the American Chemical Society* 132(40): 13978–13980.

Wang, Huaping, Xu-Bing Li, Lei Gao, Hao-Lin Wu, Jie Yang, Le Cai, Tian-Bao Ma, Chen-Ho Tung, Li-Zhu Wu, and Gui Yu. 2018. "Three-dimensional graphene networks with abundant sharp edge sites for efficient electrocatalytic hydrogen evolution." *Angewandte Chemie* 130(1): 198–203.

Wang, Kangpeng, Jun Wang, Jintai Fan, Mustafa Lotya, Arlene O'Neill, Daniel Fox, Yanyan Feng, Xiaoyan Zhang, Benxue Jiang, and Quanzhong Zhao. 2013. "Ultrafast saturable absorption of two-dimensional MoS2 nanosheets." *ACS Nano* 7(10): 9260–9267.

Wang, Qing Hua, Kourosh Kalantar-Zadeh, Andras Kis, Jonathan N. Coleman, and Michael S. Strano. 2012. "Electronics and optoelectronics of two-dimensional transition metal dichalcogenides." *Nature Nanotechnology* 7(11): 699.

Wen, Min, Danni Liu, Yihong Kang, Jiahong Wang, Hao Huang, Jia Li, Paul K. Chu, and Xue-Feng Yu. 2019. "Synthesis of high-quality black phosphorus sponges for all-solid-state supercapacitors." *Materials Horizons* 6(1): 176–181.

Weng, Qunhong, Xuebin Wang, Chunyi Zhi, Yoshio Bando, and Dmitri Golberg. 2013. "Boron nitride porous microbelts for hydrogen storage." *ACS Nano* 7(2): 1558–1565.

Weng, Qunhong, Xuebin Wang, Yoshio Bando, and Dmitri Golberg. 2014. "One-step template-free synthesis of highly porous boron nitride microsponges for hydrogen storage." *Advanced Energy Materials* 4(7): 1301525.Worsley, Marcus A., Peter J. Pauzauskie, Tammy Y. Olson, Juergen Biener, Joe H. Satcher Jr, and Theodore F. Baumann. 2010. "Synthesis of graphene aerogel with high electrical conductivity." *Journal of the American Chemical Society* 132(40): 14067–14069.

Wu, Tong, Jinchen Fan, Qiaoxia Li, Penghui Shi, Qunjie Xu, and Yulin Min. 2018. "Palladium nanoparticles anchored on anatase titanium dioxide-black phosphorus hybrids with heterointerfaces: highly electroactive and durable catalysts for ethanol electrooxidation." *Advanced Energy Materials* 8(1): 1701799.

Wu, Zhong-Shuai, Andreas Winter, Long Chen, Yi Sun, Andrey Turchanin, Xinliang Feng, and Klaus Müllen. 2012. "Three-dimensional nitrogen and boron co-doped graphene for high-performance all-solid-state supercapacitors." *Advanced Materials* 24(37): 5130–5135.

Wu, Zhuangzhi, Baizeng Fang, Arman Bonakdarpour, Aokui Sun, David P. Wilkinson, and Dezhi Wang. 2012. "WS2 nanosheets as a highly efficient electrocatalyst for hydrogen evolution reaction." *Applied Catalysis B: Environmental* 125: 59–66.

Xie, Keyu, and Bingqing Wei. 2014. "Materials and structures for stretchable energy storage and conversion devices." *Advanced Materials* 26(22): 3592–3617.

Xing, Chenyang, Guanghui Jing, Xin Liang, Meng Qiu, Zhongjun Li, Rui Cao, Xiaojing Li, Dianyuan Fan, and Han Zhang. 2017. "Graphene oxide/black phosphorus nanoflake aerogels with robust thermo-stability and significantly enhanced photothermal properties in air." *Nanoscale* 9(24): 8096–8101.

Xing, Chenyang, Shiyou Chen, Meng Qiu, Xin Liang, Quan Liu, Qingshuang Zou, Zhongjun Li, Zhongjian Xie, Dou Wang, and Biqin Dong. 2018. "Conceptually novel black phosphorus/cellulose hydrogels as promising photothermal agents for effective cancer therapy." *Advanced Healthcare Materials* 7(7): 1701510.

Xu, Mingsheng, Tao Liang, Minmin Shi, and Hongzheng Chen. 2013. "Graphene-like two-dimensional materials." *Chemical Reviews* 113(5): 3766–3798.

Xu, Ping, Bingqing Wei, Zeyuan Cao, Jie Zheng, Ke Gong, Faxue Li, Jianyong Yu, Qingwen Li, Weibang Lu, and Joon-Hyung Byun. 2015. "Stretchable wire-shaped asymmetric supercapacitors based on pristine and MnO2 coated carbon nanotube fibers." *ACS Nano* 9(6): 6088–6096.

Xu, Yuxi, Kaixuan Sheng, Chun Li, and Gaoquan Shi. 2010. "Self-assembled graphene hydrogel via a one-step hydrothermal process." *ACS Nano* 4(7): 4324–4330.

Xu, Yuxi, Qiong Wu, Yiqing Sun, Hua Bai, and Gaoquan Shi. 2010. "Three-dimensional self-assembly of graphene oxide and DNA into multifunctional hydrogels." *ACS Nano* 4(12): 7358–7362.

Yang, Bowen, Junhui Yin, Yu Chen, Shanshan Pan, Heliang Yao, Youshui Gao, and Jianlin Shi. 2018. "2D-black-phosphorus-reinforced 3D-printed scaffolds: a stepwise counter-measure for osteosarcoma." *Advanced Materials* 30(10): 1705611.

Yang, Chen, Jinfeng Wang, Ying Chen, Dan Liu, Shaoming Huang, and Weiwei Lei. 2018. "One-step template-free synthesis of 3D functionalized flower-like boron nitride nanosheets for NH3 and CO$_2$ adsorption." *Nanoscale* 10(23): 10979–10985.

Yang, Jieun, Damien Voiry, Seong Joon Ahn, Dongwoo Kang, Ah Young Kim, Manish Chhowalla, and Hyeon Suk Shin. 2013. "Two-dimensional hybrid nanosheets of tungsten disulfide and reduced graphene oxide as catalysts for enhanced hydrogen evolution." *Angewandte Chemie International Edition* 52(51): 13751–13754.

Yang, Kai, Shuai Zhang, Guoxin Zhang, Xiaoming Sun, Shuit-Tong Lee, and Zhuang Liu. 2010. "Graphene in mice: ultrahigh in vivo tumor uptake and efficient photothermal therapy." *Nano Letters* 10(9): 3318–3323.

Yang, Kai, Jianmei Wan, Shuai Zhang, Bo Tian, Youjiu Zhang, and Zhuang Liu. 2012. "The influence of surface chemistry and size of nanoscale graphene oxide on photothermal therapy of cancer using ultra-low laser power." *Biomaterials* 33(7): 2206–2214.

Yasaei, Poya, Bijandra Kumar, Tara Foroozan, Canhui Wang, Mohammad Asadi, David Tuschel, J. Ernesto Indacochea, Robert F. Klie, and Amin Salehi-Khojin. 2015. "High-quality black phosphorus atomic layers by liquid-phase exfoliation." *Advanced Materials* 27(11): 1887–1892.

Yong, Yang-Chun, Xiao-Chen Dong, Mary B. Chan-Park, Hao Song, and Peng Chen. 2012. "Macroporous and monolithic anode based on polyaniline hybridized three-dimensional graphene for high-performance microbial fuel cells." *ACS Nano* 6(3): 2394–2400.

Yu, Cunjiang, Charan Masarapu, Jiepeng Rong, Bingqing Wei, and Hanqing Jiang. 2009. "Stretchable supercapacitors based on buckled single-walled carbon-nanotube macrofilms." *Advanced Materials* 21(47): 4793–4797.

Yu, Jiali, Weibang Lu, Shaopeng Pei, Ke Gong, Liyun Wang, Linghui Meng, Yudong Huang, Joseph P. Smith, Karl S. Booksh, and Qingwen Li. 2016. "Omnidirectionally stretchable high-performance supercapacitor based on isotropic buckled carbon nanotube films." *ACS Nano* 10(5): 5204–5211.

Yu, Xiaoyun, Mathieu S. Prévot, Néstor Guijarro, and Kevin Sivula. 2015. "Self-assembled 2D WSe 2 thin films for photoelectrochemical hydrogen production." *Nature Communications* 6: 7596.

Yun, Qinbai, Qipeng Lu, Xiao Zhang, Chaoliang Tan, and Hua Zhang. 2018. "Three-dimensional architectures constructed from transition-metal dichalcogenide nanomaterials for electrochemical energy storage and conversion." *Angewandte Chemie International Edition* 57(3): 626–646.

Zaman, Muhammad H., Linda M. Trapani, Alisha L. Sieminski, Drew MacKellar, Haiyan Gong, Roger D. Kamm, Alan Wells, Douglas A. Lauffenburger, and Paul Matsudaira. 2006. "Migration of tumor cells in 3D matrices is governed by matrix stiffness along with cell-matrix adhesion and proteolysis." *Proceedings of the National Academy of Sciences* 103(29): 10889–10894.

Zhang, Jian, Shaohua Liu, Haiwei Liang, Renhao Dong, and Xinliang Feng. 2015. "Hierarchical transition-metal dichalcogenide nanosheets for enhanced electrocatalytic hydrogen evolution." *Advanced Materials* 27(45): 7426–7431.

Zhang, Jian, Qi Wang, Lianhui Wang, Xing'ao Li, and Wei Huang. 2015. "Layer-controllable WS 2-reduced graphene oxide hybrid nanosheets with high electrocatalytic activity for hydrogen evolution." *Nanoscale* 7(23): 10391–10397.

Zhang, Jintao, Zhen Li, and Xiong Wen Lou. 2017. "A freestanding selenium disulfide cathode based on cobalt disulfide-decorated multichannel carbon fibers with enhanced lithium storage performance." *Angewandte Chemie International Edition* 56(45): 14107–14112.

Zhang, Li Li, and X. S. Zhao. 2009. "Carbon-based materials as supercapacitor electrodes." *Chemical Society Reviews* 38(9): 2520–2531.

Zhang, Qiangqiang, Yu Wang, Baoqiang Zhang, Keren Zhao, Pingge He, and Boyun Huang. 2018. "3D superelastic graphene aerogel-nanosheet hybrid hierarchical nanostructures as high-performance supercapacitor electrodes." *Carbon* 127: 449–458.

Zhang, Wen, Zhouyi Guo, Deqiu Huang, Zhiming Liu, Xi Guo, and Huiqing Zhong. 2011. "Synergistic effect of chemo-photothermal therapy using PEGylated graphene oxide." *Biomaterials* 32(33): 8555–8561.

Zhang, Xuetong, Zhuyin Sui, Bin Xu, Shufang Yue, Yunjun Luo, Wanchu Zhan, and Bin Liu. 2011. "Mechanically strong and highly conductive graphene aerogel and its use as electrodes for electrochemical power sources." *Journal of Materials Chemistry* 21(18): 6494–6497.

Zhang, Zhen, Yundan Liu, Long Ren, Han Zhang, Zongyu Huang, Xiang Qi, Xiaolin Wei, and Jianxin Zhong. 2016. "Three-dimensional-networked Ni-Co-Se nanosheet/nanowire arrays on carbon cloth: a flexible electrode for efficient hydrogen evolution." *Electrochimica Acta* 200: 142–151.

Zhang, Zhitao, Meng Liao, Huiqing Lou, Yajie Hu, Xuemei Sun, and Huisheng Peng. 2018. "Conjugated polymers for flexible energy harvesting and storage." *Advanced Materials* 30(13): 1704261.

Zhao, Jinping, Wencai Ren, and Hui-Ming Cheng. 2012. "Graphene sponge for efficient and repeatable adsorption and desorption of water contaminations." *Journal of Materials Chemistry* 22(38): 20197–20202.

Zhao, Yang, Chuangang Hu, Yue Hu, Huhu Cheng, Gaoquan Shi, and Liangti Qu. 2012. "A versatile, ultralight, nitrogen-doped graphene framework." *Angewandte Chemie International Edition* 51(45): 11371–11375.

Zheng, Xin, Yangwei Zhang, Runze Chen, Zhongjie Xu, and Tian Jiang. 2015. "Z-scan measurement of the nonlinear refractive index of monolayer WS 2." *Optics Express* 23(12): 15616–15623.

Zhu, Minshen, Yang Huang, Yan Huang, Hongfei Li, Zifeng Wang, Zengxia Pei, Qi Xue, Huiyuan Geng, and Chunyi Zhi. 2017. "A highly durable, transferable, and substrate-versatile high-performance all-polymer micro-supercapacitor with plug-and-play function." *Advanced Materials* 29(16): 1605137.

6 Emerging Applications of 3D Structures of BP

6.1 INTRODUCTION

The treatment of black phosphorus (BP) nanosheets as the basic unit for the assembly of a 3D construction might be a smart technique for achieving its extraordinary properties (Yun et al. 2018; Wang, Li, et al. 2018; Zhang, Li, and Lou 2017; Zhang et al. 2018). Conversely, some characteristics, such as the compassion of BP nanosheet (to water, air, and oxidants), make the construction of 3D BP difficult (Wen et al. 2019). BP has puckered geometry (which is favorable for energy storage and conversion applications) and a large surface area, strong mechanical strength (E94 GPA), and high value of carrier mobility value (10 000 cm^2/Vs) (Wu, Hui, and Hui 2018; Qiao et al. 2014; Wei and Peng 2014; Zhu, Zhang, et al. 2017; Zhu, Sun, et al. 2018; Sajedi-Moghaddam et al. 2017). For example, BP has a theoretical value of specific capacity of around 2596 mAh g1, which is about seven times more than graphite in lithium-ion applications and sodium-ion (Na$^+$) batteries (Sun et al. 2014). The specific capacity yield in case of BP nanosheets has been confirmed as 1968 mAh g1 at 100 mA g1 when working as an anode in sodium-ion batteries. and the supercapacitor based on BP nanosheets has a capacitance value of around 45.8 F g1 at 10 mV s1 (Huang et al. 2017; Hao et al. 2016). All the BP sponge-based electrodes for supercapacitors have excellent stability as well as a maximum value of specific capacitance with a limit of 80 F g1 at 10 mV s1. It is highly desired to develop a 3D BP architecture, as the 3D structures of 2D BP nanomaterial is expected to perform better for practical industrial applications in the field of potential and biomedical applications. During the past five years, a lot of research work has been done, but there are still many fundamental issues yet to be resolved. BP has been utilized already in different applications, many of which, including transistors (Wu, Hui, and Hui 2018), optoelectronics (Viti et al. 2015), batteries (Zhou et al. 2017), and biomedical applications, are discussed in this chapter (Xing et al. 2018). This chapter discusses the recent evolution of BP-based heterostructures, as well as a wide array of applications, including mechanical, optoelectronic, energy storage, photocatalysis, electrocatalysis, and biological applications.

DOI: 10.1201/9781003217145-6

6.2 3D ARCHITECTURE OF BP IN OPTOELECTRONICS

6.2.1 PHOTO-DETECTORS

Due to the narrow bandgap (limit from ~2 eV to 0.33 eV), the layer thickness value of the film gradually increases from a single layer to tens of nm; due to the presence of such properties, 2D BP is a very reliable and attractive candidate in electronics and photonic devices (Li et al. 2014; Das et al. 2014; Ren, Li, et al. 2017). In the near- and mid-IR regimes, BP is familiar due to its worthy sensitivity. Several electronics and photonics devices have been established, such as polarization-sensitive, (Yuan et al. 2015) multispectral imager, (Engel, Steiner, and Avouris 2014) mid-IR photodetectors (Guo, Pospischil, et al. 2016), and waveguide integrated near-IR, etc. (Youngblood et al. 2015). Also, on excitation in both cases, either in case of incident of normal light or a large device footmark, these developed devices usually suffer from quite low value of responsivity, i.e., response time.

Chen et al. (2017) demonstrate the incorporation of a BP-based photodetector in a 3D hybrid structure of silicon-based nanoplasmonics and photonics structures, and the network is illustrated conceptually in Figure 6.1a. There are three different layers designed and constructed on a standard stand of silicon-on-insulator (SOI). According to the labeling, the lowest coating layer (consisting of silicon photonic waveguide phenomenon), has the low transmission damage; the middle coated thin layer (containing a metal grating and a nanogap) is related to plasmonic, and the top layer (having a direct connection with the metallic nanogap) contains exfoliated BP nanoflakes (Chen et al. 2017).

A grid underlying the plasmonic structure is modeled on the waveguide to open light out of the plane via a distance layer, which connects light from the waveguide to the nanogap. The waveguide grating's output light (basic TE waveguide mode) is polarized in the x-direction, which is perpendicular to the nanogap direction. The metal grating in the plasmon structure transforms emissions from the waveguide to surface plasmon polariton (SPP) wave and concentrates it into a nanogap where the optical intensity is greatly increased. Figure 6.1b illustrates a schematic cross-sectional image of this structure superimposed with the electric field amplitude of the optical mode simulated by the 2D finite-difference time-domain (FDTD) approach (normalized to the source amplitude |E|/|E0|) (Chen et al. 2017).

There is an improved electric field amplitude of the averaged assessment of optical intensity in the sheet of BP (panel on right) and the profile inside the nanogap (panel on left) as shown in Figure 6.1c, which strongly recommended that the augmentation factor range from 20 to 45 as compared to the output intensity values of waveguide grating. Moreover, by a gap size in the range of 60 nm, a graphical representation (plasmonics nanostructure material having a transmission spectrum obtained as a result of the simulation) is shown in Figure 6.1d. Plots of the specific involvement (a combination of gold and the BP) originate in the simulation process, and the total value of the optical absorption spectrum (hybrid BP nanogap) is mentioned in Figure 6.1e. Although the BP contributes less absorption (13 percent at peak) than metal and in another ward, the light engaged by BP nanoflakes is near around ~11% at peak (controlled to the power value discharged by the source of waveguide-based gratings). Compared to other reported methods, the width of the designed aperture (60 nm) is much smaller (Guo, Pospischil,

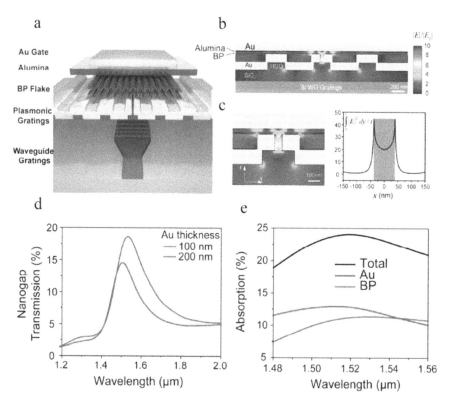

FIGURE 6.1 The hybrid 3D structure of BP and silicon photonics, nanoplasmonics. a) Design of the incorporated hybrid method for each layer with proper labels. b) The structure overlapped standardized to the foundation amplitude |E|/|E0| (with electric field amplitude) of the optical mode as a result of simulation by the method named as 2D finite-difference-time-domain (FDTD) and the cross-sectional view of the mentioned geometry. c) The result explained an improved electric field amplitude of the averaged value of optical-based value of intensity in the structure of BP-based layer (panel on right) and the profile inside the nanogap (panel on left). d) The simulated transmission spectrum with a gap thickness in the range of 60 nm, the graphical representation (plasmonics nanostructure having the transmission spectrum obtained as a result of simulation). e) Plots of the specific involvement (combination of BP and the gold) originate in simulation process and the total value of optical absorption spectrum (nanogap-BP hybrid), with contributions from the metal and the BP. Adapted with permission from Chen et al. (2017), Copyright © 2017, American Chemical Society.

et al. 2016; Youngblood et al. 2015). Figure 6.2a shows the designing of a fabricated device captured by the optical microscope image. For monitoring the surface morphology, a dry transfer method was used for the fabrication of thin-film from exfoliated BP nanoflakes (Castellanos-Gomez et al. 2014). Figure 6.2b is shown in the atomic force microscope (AFM) image (the surface morphology of BP-nanogap area) earlier incorporating the top gate. The BP flake (having the thickness around 20 nm and width of 10 μm) has been magnificently moved as well as incorporated with the plasmonic

a   b

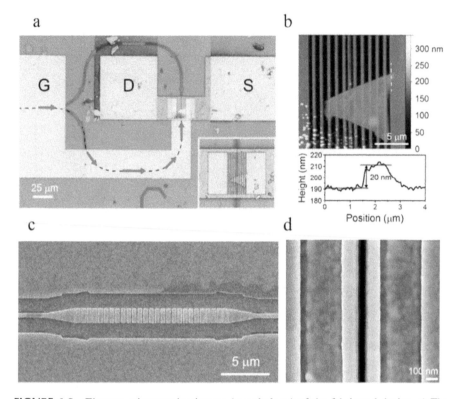

FIGURE 6.2 Electron microscopic pictures (morphology) of the fabricated device. a) The designing of a fabricated device captured by the optical microscope image. For monitoring the surface morphology, a method named as dry transfer was used for the construction of thin-film from exfoliated BP nanoflakes. b) AFM image shows the size/shape and roughness of the BP-nanogap region having thickness around 20 nm at lower region of image. c) SEM images shows the details of the nanogap and the surface morphology of the silicon waveguide grating. d) The nanogap having a charming and attractive shape (with width of around 60 nm size) which was imperfect and also there is a deficiency of the resolution by the focused ion beam (FIB) arrangement is shown in the picture of SEM. Adapted with permission from Chen et al. (2017), Copyright © 2017, American Chemical Society.

nanostructures through the nanogap. The nanogap has a charming and attractive shape (with a width of around 60 nm), which was imperfect. Also, the system's resolution is a deficiency such as a focused ion beam (FIB) shown in the scanning electron micro-scope (SEM) picture in Figure 6.2d. The silicon-based waveguide grating (having a period of 590 nm with a duty cycle of around 78 percent) is shown in SEM images (Figure 6.2c). Additionally, it is important to note the basic mechanistic approach to how the light is transmitted from the BP nanoflake-based photodetector in the designed hybrid arrangement. In the case of using the adiabatic process (occurring without loss or gain of heat) for thinning the duty cycles to moderate the reverse reflection, the waveguide grating is designed with an efficiency of 48 percent (Figure 6.2c). Both the characteristics—silicon-based photonics and the capability of the light to transport over a wide range of distance (plasmonics capability to deliberate light and with a low power dissipation damage) on the boundary of sub-diffraction level—of the hybrid system significantly benefit the performance of the photodetector.

A short channel photodetector is made (formed as a result of direct addition of BP on the plasmonic nanogap which supporting built-in photoconductive improvement with a high recognition value and limited bandwidth) and the photoresponse behavior of the device is slowed in that case when it utilities in the photoconductive style having a functional source-drain bias voltage.

The photoresponse of the device was tested when it was operating in the photoconductive mode with a source-drain bias applied. Figure 6.3a. illustrates the source-drain current IDS at different optical power levels, denoted in the legends as the power emitted by the waveguide grating, with a fixed VG of 8V. The magnitude of photocurrent created in the BP channel grows with the optical power. Figure 6.3b

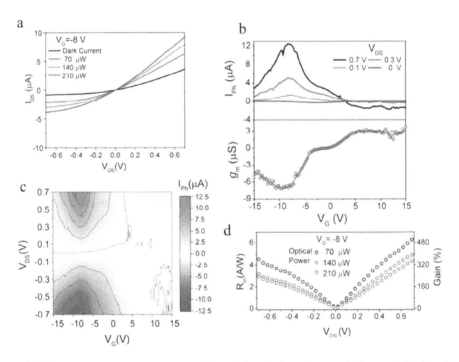

FIGURE 6.3 Photoresponse property of the BP based photodetector. a) The graph is plotted between the source-drain bias VDS against the source-drain value of current (IDS) mentioned for the BP FET (at a static quantity of gate voltage VG ~(−8V)) under a variable value of optical power levels. b) The interaction between gate voltage and the photocurrent of the device, and it is shown that the maximum value of photocurrent arises adjacent a gate bias voltage of around −8 V. c) The photocurrent (2D curve plot) strongly focuses on the bias and the gate voltage. d) The photocurrent measurement results (Iph), the intrinsic responsivity value (RIn) for various kind of optical powers and analyzed that the engaged (value of optical power in BP) is as high as 6.25 A/W (conforming to a photoconductive enhancement of around 500 percent). Adapted with permission from Chen et al. (2017), Copyright © 2017, American Chemical Society.

shows the interaction between gate voltage and the photocurrent of the device and it is shown that the maximum value of photocurrent arises near a gate bias voltage of around −8 V (Chen et al. 2017). The photocurrent (indicating an internal gain mechanism) corresponds with the peak value of the field-effect transistor FET transconductance g_m, as mentioned in the lower rectangle of Figure 6.3b. The 2D curve of the photocurrent strongly relies on the bias and the gate voltage, shown in Figure 6.3c, and it is found that in the highly doped regions there is a very low quantity of device negative photocurrent (Youngblood et al. 2015; Freitag et al. 2013) and a large metal area in the fabricated device might be responsible for this deficiency, the low negative photocurrent (well-organized heat basin for the BP channel) is responsible for the dissipation of heat quickly. Depends on the photocurrent measurement results (Iph), the intrinsic responsivity value (RIn) for various optical powers is designed and plotted in Figure 6.3d.

The external value for the photodetector's responsivity is measured as 214 mA/ W and it is possible only in the case that the expected optical-based emission power value from the BP enclosed around 18 percent from the net grating area is around nearly 12.6 μW.

This integration strategy trusts the benefits of high field limitation of a plasmonic-based nanogap, the narrow value of bandgap for BP as well as the loss of the low value of propagation for silicon waveguides to achieve high responsivity; the photodetector showed an intrinsic responsivity in the upper limit of 10 A/W yielded an internal gain pathway with 3-dB roll-off frequency of ~150 MHz.

The developed demonstration of a photodetector with fast frequency response and high responsivity has only been seen in the first step. In summary, the integrated and updated methodology can be further extended for the preparation of well-organized optical-based modulators (Melikyan et al. 2014), co-integration of 2D electronics with optoelectronics (Fiori et al. 2014), nonlinear-based optical devices (Youngblood et al. 2016), and with various types of potential applications (Willner et al. 2013).

6.2.2 BIOMEDICAL-BASED APPLICATIONS

Although the existence of phosphorus (P) as a vital component in the human body, responsible for developing approximately ≈1% of total body weight in the form of bone component, is scientifically proven, there remains a shortage of research focusing on BP nanosheets and the nano-bio interactions (with the body fluid (liquid plus gas) environment especially a biological trigger for osseointegration and osteogenesis).

The presence of many fascinating properties (biocompatibility, layer-dependent bandgap, moderate carrier mobility, sizeable surface-area-to-volume ratio, and biodegradability) makes the 2D BP an active and reliable candidate for biomedical-based applications, including photodynamic therapy (PDT), photothermal therapy (PTT), 3D printing, drug delivery, bioimaging, theragnostics, and biosensing. To enhance the efficiency of BP, various modification strategies such as doping and surface chemical modification have been successfully employed during the past decade (Zheng et al. 2018). Although 2D BP is still facing various unresolved problems in the reproducible synthetic process, stability, and toxicity, it is also considered a unique candidate for various biomedical and potential applications. The abundant phosphate on the BP contents' surface site tends to combine physically and chemically with various proteins, drugs, nucleic acid, and even various nanoparticles to provide an adequate performance. The various structures of BP can also combine with various metal ions required for the human body through the coordination linkage, and the resultant hybrid product can be treated against various diseases. Moreover, BP-based structures have been prepared in BP heterojunction and BP gel, increasing the physical and chemical properties and making it a reliable candidate for biomedical applications (Wang, Huang, et al. 2018).

Combination therapy for cancer has also been considered in recent years to suppress solid tumors, thereby disseminating tumors and metastatic tumors. Near-infrared irradiation cannot cause PPTs for disseminated tumors and metastatic tumors (Nam et al. 2018). To achieve successful chemo-photoimmunotherapy of colorectal cancer, Ou et al. (2018) used BP nanosheets to load drugs, targeting agents, and cancer growth inhibitors. Improved therapeutic results are supposed to be achieved by a combination of BP and immunity. Huang et al. (2018), introduced BP-based wound dressings, protecting people against bacterial infection and encouraging wound repair.

It is suggested that BP be considered an active candidate for clinical research and application in the near future. However, BP still faces some unanswered issues about preparation processes, stability, and toxicity. As scientists/researchers from the biomedical, pharmaceutical, and even industrial sectors are increasing their involvement, 2D BP can be further explored in the biomedical application after developing the 3D structures of BP (Luo et al. 2019).

Due to various unique physicochemical characteristics (in biomedical applications including excellent photothermal performance and tissue engineering), nowadays, BP nanosheets are considered a newly emerging 2D layer-structured material and have adopted a wide range of attention for biological applications (Chen et al. 2015; Sun et al. 2015; Lv et al. 2016; Sun et al. 2017). The widely used BP in photothermal therapy and tissue engineering is discussed in the following sections.

6.2.2.1 Photothermal Performance

The essential instability of BP nanosheets in the physiological environment and vascular system is a problem when considering its future clinical translation. As a result of interactions of BP with, oxygen, visible light, and water, there is a process of degradation of BP and oxidation in aqueous media and, as a result, phosphate is the major product, which might help the subsequent low therapeutic outcome and lessened photothermal performance (Shao et al. 2016; Zhao et al. 2016; Walia et al. 2016; Tao, Ji, et al. 2017; Qiu et al. 2018)

As a result of *in situ* degradation of BP nanosheets, there might be a chance of transformation into P-based agents that might help enhance the osteosarcoma therapeutics and the bone regeneration process due to the exceptional physiochemical-like property of 2D BP nanosheets (Walia et al. 2016). In order to formulate a BP-BG based bifunctional scaffold for the subsequent bone regeneration and photothermal therapy of osteosarcoma recently Yang et al. (2018) developed an exquisite therapeutic and a novel constructed platform as a result of participating 2D BP nanosheets extended

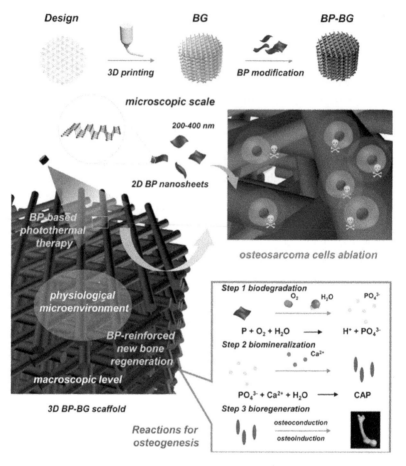

FIGURE 6.4 The fabrication process for an exquisite therapeutic and a novel constructed BP-BG scaffold platform as a result of participating 2D BP nanosheets extended toward a 3D printed bioglass (BG)-based scaffold material. Adapted with permission from Yang et al. (2018), Copyright © 2018, WILEY–VCH Verlag GmbH & Co. KGaA, Weinheim.

toward a 3D printed bioglass (BG)-based scaffold material (Figure 6.4). There is an urgent need to develop stepwise treatment theory for minimizing the aforesaid restriction of recent working therapeutic tactics for osteosarcoma.

1. Photothermal therapy (the excellent photothermal presentation of BP-based nanosheets was employed) after surrounded by bone faults resulting from surgical resection (surgical removal of part of a structure or organ).
2. 3D-printed BG-based scaffolds play a key role in achieving cell propagation, diversity, the formation of new blood vessels, and the vascularization process, due to excellent properties for osteoconduction, osteoinduction, and ontogenesis (Li and Liu 2017; Wang et al. 2016).

3. There is a process of scaffold degradation gradually (as new bone tissue components) as a result of the restoration of pathological-changed areas. The bifunctional BG scaffold material functions as a medium for supporting the BP nanosheets as well as for the repairing the tumor-induced bone defects, and an outstanding *in situ* biomineralization performance to facilitate osseointegration process.

As a result of the oxidation process of BP nanosheets there is an increase in the release of phosphate ($PO4^{3-}$) ion; $PO4^{3-}$ is using anionic ligands to encourage the development of new calcium phosphate (CAP) nanoparticles (NPs), it is also used to remove calcium ions (Ca^{2+}) from the physiological atmosphere. The as-designed 3D scaffold (BP-BG) shows outstanding presentation for photothermal ablation of tumors, and this is possible due to the high photothermal-transformation efficiency of BP nanosheets. Figure 6.5a shows the increased temperature from 32.4 to 68.7°C of BP-BG scaffold (quantity of BP = 200 ppm, the power density value is 1.0 W/ cm²) under laser irradiation 808 nm for a time duration of five minutes. BP nanosheets perform a leading and critical role in photothermal therapy (PTT) while the other constituents are unable to perform and comparatively, in the presence of pure BG scaffold, the photothermal effect is negligible. Moreover, there is a direct relationship between the soaking time and temperature, which is measured and represented in Figure 6.5b.

The cytotoxicity of BP-BG scaffolds is deliberated mainly to demonstrating the cell-level photothermal therapy. As a result of 3 d incubation, the osteosarcoma cells (Saos-2) (cancerous tumor in a bone) follow and flourish on BP-BG scaffolds; this evidence strongly suggests the BP-BG based scaffolds have excellent cytocompatibility. Additionally, Saos-2 cells on the surface of BP-BG-based structure of scaffolds material are more than those protected on BG-based scaffolds material; this might be due to the BP constituent assisting as friendly bioactive sites to smooth the cell propagation process. The relatively high value of (PTT) effectiveness of BP-BG-based scaffolds-based structural material is analyzed *in vitro* due to the opportunity of Saos-2 cells after interacting with the BP-BG-based structure of scaffold.

It was analyzed that after the treatment of NIR irradiation, the BP-BG-based structure scaffold was considerably reduced (simple BG, a combination of BP-BG and NIR treated groups). This evidence strongly recommended the feasibility of BP-BG scaffold for *in vivo* oncological application. Figure 6.5c shows the irradiation time and possessions of laser power density on photothermal performance (PTP) for carefully judging the application outlook of BP-BG scaffolds (Yang et al. 2018).

Figure 6.5d shows the live/dead cells observed on well-ordered tube-shaped scaffolds, demonstrating that a well-controlled interconnected macroporous structure is invented due to 3D printing performance and the confocal laser scanning microscope (CLSM). There is a sharp divergence (in both cases before and after photothermal therapy) that demonstrates the presence of different (such as red or green) fluorescence colors emitted from the analyzed live/dead cells as well as also demonstrating the 3D hierarchical and well-ordered alignment of BP-BG scaffolds material. Additionally, the tumor-treated symbolic mice were distributed into four

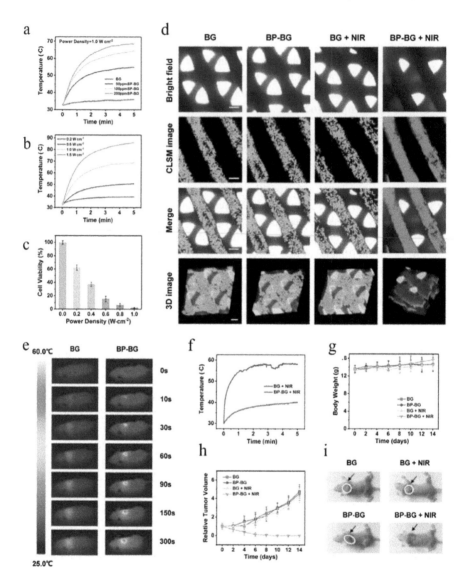

FIGURE 6.5 *In vivo* and *in vitro* photothermal-based efficiency evaluation of combined BP-BG scaffolds. a) *In vitro*-based photothermal increased in temperature from 32.4 to 68.7°C of BP-BG scaffold (quantity of BP = 200 ppm, the power density value is 1.0 W/ cm²) under laser irradiation 808 nm for a time duration of five minutes. b) At various laser power densities of NIR, there is a direct relationship between the soaking time and temperature measured and represented. c) The irradiation time and possessions of laser power density on photothermal performance (PTP) for carefully judging the application outlook of BP-BG scaffolds. d) The live/dead cells observing on well-ordered tube-shaped scaffolds, demonstrating the well-controlled interconnected macroporous structure invented due to 3D printing performance and confocal laser the scanning microscope (CLSM). There is a sharp divergence (in both cases before and after photothermal therapy) demonstrating the presence of different (such

FIGURE 6.5 Continued

as red or green) fluorescence colors emitted from the analyzed live/dead cells as well as also demonstrating the 3D hierarchical and well-ordered alignment of BP-BG scaffolds material. e) After the process of NIR irradiation, the BP-BG scaffolds kept in the tumor cell are producing durable and strong hyperthermia for tumor excision. f) When the monitoring time is increased to five minutes, the temperature is drastically increased to 58°C in osteosarcoma tissue consistent too. g) After different trial the group such as (simple BG, combination of BP-BG, BG + NIR irradiation process), the interaction of time on body weight turns of bare mice (mean ± SD, n = 7), BP-BG + NIR radiation therapy. h&i) After different growth of tumor curves of the mice were analyzed. It was observed that without return in the statement and consideration period of 14 days, the mice of the BP-BG-based scaffold + NIR group having the tumors were totally removed. Adapted with permission from Yang et al. (2018), Copyright © 2018, WILEY–VCH Verlag GmbH & Co. KGaA, Weinheim.

different categories: simple BG, BG + NIR laser, combination of BP-BG, BP-BG + NIR laser, etc. After a skin scratch was prepared at the tumor's edge, the structure of scaffold-based material was bounded into the tumor-bearing cell's main center location. Moreover, after NIR irradiation, the BP-BG scaffold was kept in tumor tissue, producing durable hyperthermia for tumor excision, as shown in Figure 6.5e.

It is clearly observed that after the passage of time (one minute), there is a gradual increment in temperature of tumor tissue from 30 to 55°C. In addition, when the monitoring time is increased to five minutes, the temperature is drastically increased to 58°C. Comparatively, there is a deficiency of substantial temperature proliferation (power intensity: 1 W cm^{-2}) in both cases showed by the BG scaffold independently, as shown in Figure 6.5f. It is also important to record the tumor volumes and increase or decrease body weights of the tumor-treated bare mice to estimate the therapeutic efficiency and biocompatibility of BP-BG scaffold material for the *in vivo* monitoring process. It was found that all the mice under consideration displayed insignificant weight variations.

This evidence strongly confirms irrelevant opposing possessions of these above-mentioned managements on the fitness care treatment of mice (Figure 6.5g) and especially, without return in the comment and consideration period of 14 days, it is observed that the tumors cell treated for mice of the category (combined BP-BG based scaffold + NIR group) were fully removed (Figure 6.5h–i) (Yang et al. 2018; Xie et al. 2016). In comparison, during the 14-day period, a rapid increase in the volumes of tumors in groups (such as simple BG, combined BP-BG, and BG + NIR) was detected. This evidence strongly suggested that for the localized photothermal-based therapy of cancer, the BP-BG scaffold is highly biocompatible and can be used for practical application on tumor-affected bodies with a similar immune system, like mice.

In another study, Xing et al. (2018) demonstrate that cellulose and BP nanosheet (BPNS)-based green and injectable composite hydrogels have great potential for PTT against cancer (Sun et al. 2015; Sun et al. 2016). PTT against cancer was explored after a confirmation report on the nontoxic behavior and biological and genetic safety of cellulose/BPNS-based hydrogels material. After NIR light irradiation, the cellulose/BPNS hydrogels' *in vitro* cell killing effectiveness and *in vivo* tumor regression efficiency were assessed. After using cancer cells such as SMMC-7721, B16, and J774A.1 as prototypical cell lines, the comparative cell feasibility was determined after proper development was made with the cellulose functionalized BPNS composite hydrogels in the form of different concentration dose (0 ppm, 95 ppm, 190 ppm, 285 ppm, and 380 ppm) monitored by proper NIR irradiation treatment (1 W / cm, 10 min).

It was analyzed that the cellulose hydrogels (containing the contents of BP (of 95 and 190 ppm)) were unable to display unsatisfactory photothermal properties to kill tumor cells as shown in Figure 6.6a, in comparison to those prepared without the addition of BP. It was found that the at 285 ppm of BP the cell-killing efficiency reached to 40–60 percent and almost 100 percent when the concentration was increased to 380 ppm due to increasing BP loading levels; due to different survival abilities under heat stress, differences might be caused between the cell lines. The fluorescence images showed similar results, as shown in Figure 6.6b.

The identification of whether cells are alive or dead is based on the color identification, and it was found that in case of breathing SMMC-7721 cells there appeared a green color (stained by calcein acetoxymethyl), while in case of dead ones there was a red color (stained by propidium iodide). It was found that there is a great interaction between the concentrations of BP on the death of the cell. It was found that in the case of cellulose-based hydrogels material incubated with the SMMC-7721 cells by NIR light treatment, the cell inaugurated to die and were fully observed as dead as a result of an increase in the quantity of BP (raised from 285 to 380 ppm). The findings as mentioned above strongly recommended that the proper usage of composite hydrogels (cellulose/BPNS) for PTT required light-to-heat transformation productivity against *in vivo* cancer therapy possesses. In order to investigate *in vivo* cancer, PTT of (cellulose/BPNS composite hydrogels) human hepatocellular carcinoma representations was recognized into six-week longstanding female BALB/ c bare mice by intravenous injection filled with SMMC-7721 cells (4 × 106 cells treated for every individual mouse). There is a division of mice into four groups (five mice in each group) when the tumor volumes extended 100–200 mm^3 and the detailed about the groups (G) are mentioned as G1, G2, G3, and G4. G1 processed on cellulose-based hydrogel treatment, G2 proficient cellulose/BPNS hydrogel behavior with concentration (380 ppm), G3 underwent NIR irradiation plus cellulose hydrogel treatment, and G4 underwent cellulose/BPNS hydrogel with concentration (380 ppm) plus NIR irradiation management.

Moreover, the variation in temperature variation due to the NIR-irradiated was constantly detected with an infrared thermal detector. In G3 when the cellulose hydrogel treated with the tumor-bearing mice as shown in Figure 6.6c–d, with a slow speed, there was a slightly increased temperature of the tumor site as a result of NIR irradiation treatment (41.2 °C) (Xing et al. 2018). In the case of G4, when interacting with composite cellulose functionalized BPNS hydrogel (with 380 ppm of BP quantity), in the initial time there was a rapid growth in speed, the temperature reached 64.1°C. The tumor volume of each mouse was measured continuously after the treatments, starting from treatment day. It was found that even when the

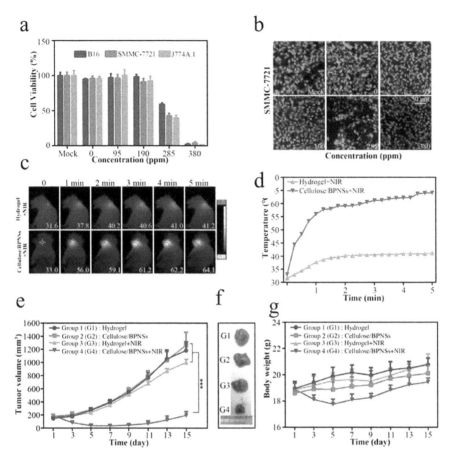

FIGURE 6.6 Photothermal (PTT) cancer therapy. a) Cellulose hydrogels (containing the contents of BP (of 95 and 190 ppm)) are unable to display unsatisfactory photothermal properties to kill tumor cells. b) There is a treatment of SMMC-7721 cells in first stage of processing like a), after that having therapy for fluorescence visualization which confirm the life cycle of cells. c&d) The variation in temperature variation due to the NIR-irradiated technique was constantly analyzed with an infrared thermal detector. In G3 when the cellulose hydrogel is used on the tumor-bearing mice with a slow speed, there is a slightly increased temperature of the tumor site due to NIR irradiation treatment (41.2°C). e) The time-dependent difference in tumor volume after the action with composite-based hydrogel (cellulose functionalized BPNSs), cellulose/BPNSs + NIR and hydrogel + NIR irradiation. The tumor volume of Group 4 (G4) is compared with the rest of all the remaining groups. f) Isolation of symbolic tumors cell from the four groups such as G1, G2, G3, and G4. g) During the passage of time there is a variation in body weight of the mice in G1, G2, G3, and G4. Adapted with permission from Xing et al. (2018), Copyright © 2018, WILEY–VCH Verlag GmbH & Co. KGaA, Weinheim.

concentration is extending to 380 ppm, there is no measurable effect on tumor growth when the cellulose hydrogel (G1) is compared, after making sure of the presence of BPNSs into the cellulose-based hydrogel (G2) system. After NIR irradiation (G4), the tumor (which was almost totally killed) exhibited significant regression. There is a slightly decreased tumor volume in G3 (cellulose hydrogel with NIR irradiation), which might be due to the inhibition of tumor growth by a nonlethal temperature increase. The variation in tumor volume with time from day 1–15 and its detail is shown in Figure 6.6e and the symbolic tumors in each group after there is an increase in time (on day 15) of mice is mentioned in Figure 6.6f. Moreover, Figure 6.6g shows the detail of the mice's body weight measurement during the treatment period (days 1–15). It was found that due to the presence of BP, BPNS-combined cellulose hydrogels have excellent photothermal and good mechanical properties as a result of their 3D networks with multiples and substantial water, which have established biocompatibility and biosafety for both the cells as well as the structure of the organ in the living body system (Tao, Zhu, et al. 2017; Xing et al. 2018). These findings strongly recommended that the treatment of cellulose functionalized BPNS composite-based hydrogels for the cancer PTT treatment is a quite reliable and effective approach for both preventing the growth of cancer as well as killing the cancer cells.

6.2.2.2 Tissue Engineering

Due to their capability to modify tissue scaffolds' surface characteristics and thus enhance their biocompatibility and cell affinity, 2D nanostructures have appeared as a promising new research direction toward tissue engineering. Many 2D materials have been documented to increase cell adhesion and proliferation, like graphene oxide (GOs) and graphene. A newly developed 2D BP-based material has recently gained interest through its unusual mechanical and electrochemical properties in biomedical applications. Herein, the synergistic effect for the cell osteogenesis-based bone-tissue engineering of both types of 2D materials has been examined. First, BP has been wrapped in GO nanosheets that have been negatively charged and then adsorbed to 3D (three-dimensional) poly(propylene) positive-loaded scaffolds (Novoselov et al. 2005; Xia et al. 2014). Liu et al. (2019) exposed the osteogenic potential of the 3D-printed based biodegradable scaffolds material, treated for coating with 2D nanomaterials (Figure 6.7a). The practical 3D bandages comprise a related biodegradable-based PPF matrix, covered casually with 2D GO nanolayers, BP nanolayers, GO, and BP wrapped composites (GO@BP). GO was approved with its famous protein adsorption capability to increase initial cell adherence to the surface site of scaffolds (Figure 6.7b). Nanosheets of BP are chosen to examine the impact on cell proliferation, expansion, form, filament growth, and osteogenic differentiation of the effect of phosphorus discharge and increased structural complexity (Figure 6.7c).

Improvements were also explored in the material characteristics of 3D-based PPF skinning materials by each 2D nanomaterial, like phosphate release, surface roughness, and mineralization (Figure 6.7d). After the establishment process, polymer scaffolds material was built to reinforce new bones in the field (Figure 6.7e) in conjunction with biodegradation via ester bond hydrolysis over time. The desired 3D scaffolds were constructed using CADs with dimensions of 5mm length, width,

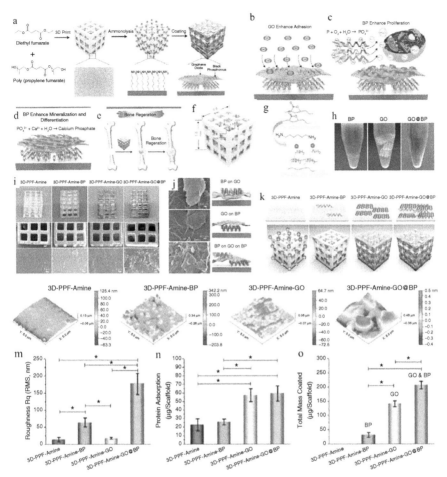

FIGURE 6.7 a) Schematic representation of the 3D printed scaffold manufacturing process, functionalized with 2D GO and BP nanosheets. b) GO nanosheets boost cell adhesion to the scaffold surface. c) BP nanosheets release phosphate groups during degradation that increase the pre-osteoblast proliferation. d) BP nanosheets release phosphates, which also boost the mineralization of bodybuilding and cell differentiation and make it an appealing choice. e) Schematic display of 2D material-functionalized 3D scaffolds for application of bone tissue engineering. f) CAD type model dimensions. g) 3D-PPF scaffold surface ammonolysis mechanisms. h) 3D scaffold covering of GO, BP, and GO@BP hybrid material. i) Photographs and SEM images of after-coating scaffolds. j) SEM images of varying BP and GO morphologies on the 3D-PPF-Amine-GO@BP scaffold surfaces. k) Schematic visualization of the four forms of 3D scaffolds. l) AFM photograph. m) Averaged roughness (n = 5; *: p <0.05) of the 3D based scaffolds material. n) Protein adsorption process and o) total mass contents coating process on the functionalized 3D scaffolds (n = 3; *: p < 0.05). Adapted with permission from Liu et al. (2019), Copyright © 2019, American Chemical Society.

and height outside dimensions (Figure 6.7f). 3D scaffold ammonolyses were performed for 30 minutes in a solution of hexamethylenediamine/isopropyl alcohol at a temperature of 60°C. The ammonolysis molecular-based mechanistic approach incorporates the polypropylene fumarate (PPF) polymer chain of amine groups in hexamethylenediamine, as shown in Figure 6.7c. The ammonolysis was performed for 30 minutes at 60°C in a hexamethylenediamine/isopropyl-based alcoholic solution. The ammonolysis molecular mechanism incorporates the PPF polymer chain of amine groups in hexamethylenediamine, as shown in Figure 6.7g. Scaffolds materials were deposited after ammonolysis in various solutions containing either a concentration value of 1 mg/mL BP nanosheets, GO nanosheets about 1 mg/mL of, or 1 mg/mL of each GO and BP, as shown in Figure 6.7h.

The photos of post-coating scaffold materials are shown in Figure 6.7i. SEM-pictures display different morphological features on the nanoscale of the surface of four scaffold forms. The surface is relatively smooth without any fragments for 3D-PPF-Amine scaffolds in the absence of the coating process. Nevertheless, after the coating with BP or GO, a deposited form BP or a thin nanolayer of GO-based nanosheets were found in 3D-PPF-Amine-BP or 3D-PPFAmine GO scaffold surfaces, respectively, each with a crust. The surfaces were covered by an abundance of BP nanolayers covered by a GO coating for the 3D-PPF-Amine-GO@BP scaffolds coated with the BP and GO nanosheets (Figure 6.7i). Detailed scoping revealed that BP and GO have been placed jointly with BP and GO layers, GO on the BP layer or BP contents on GO layer, and then onto the BP layer (Figure 6.7j). A SEM observation concluded the surface morphologies of four scaffold forms, and a schematic illustration is mentioned in Figure 6.7k.

Further evidence was collected from morphological topographies found with SEM observations with testing by AFM (Figure 6.7l). The statistically high roughness of BP-coated scaffolds material equated to GO-coated scaffolds has been demonstrated in the root-mean square rouge calculation (Figure 6.7m). Many studies have documented the critical role of surface roughness in increasing cellular adhesion and proliferation (Chung et al. 2003; Deligianni et al. 2000; Linez-Bataillon et al. 2002). Protein adsorption has substantially increased protein adsorption in nanosheets covered by GO or GO@BP compared with others in cell culture media (Figure 6.7n). This finding is compatible with previously published studies, where a reliable and robust process based on the adsorption of the proteins to the GO-nanolayers was also observed (Xiong et al. 2017; Ku and Park 2013). The increased ability to adsorb protein is mainly due to the ionic bonding ability and the vast surface area site of GO-nanomaterials (Shi et al. 2012; Li et al. 2016) (Figure 6.7o). This can be attributed to the negative GO surface charges, which could produce robust electric forces, allowing GO layers to bind better than BP to the positively charged 3D ammonium scaffold surface. The GO surface area is also broader than BP, enabling the unbinding process of GO from the scaffold, but is more complicated than BP. The GO@BP mixture utilizes the GO material's high adsorption capability while simultaneously placing BP nanosheets on a 3D scaffold surface. This combination has allowed the highest coating mass contents (Figure 6.7o), derived from the mass contents of GO and the BP nanolayers wrapped in the GO layer, to be achieved by the 3D-PPFAmine-GO@BP scaffolds (Liu et al. 2019).

EDXS elemental mapping has defined the quantity and distribution process of the BP species adsorbed on scaffolds. There is a significantly higher concentration of phosphorus species on the cross-sectional area of 3D-PPF-Amine-GO@ BP scaffolds than 3D-PPF-Amine-BP scaffolds in Figure 6.8a. The quantitative analysis calculated the high phosphorus content of a 3D-PPF-Amine GO@BP scaffold compared with 3DPPF-Amine-BP scaffolds (Figure 6.8b). This shows that the layer of GO, wrapping BP nanosheets, helped bind BP nanosheets on the scaffold surface. Further analysis was made on both the body's temperature (37°C) and room temperature to remove the 3D-PPF-Amine-GO@BP scaffold phosphates and 3D-PPF-Amine-BP. For scaffolds 3D-PPF-Amine-BP, during the time interval of the first eight days following the engagement process in deionized (DI) water, phosphate ionic species were gradually released from scaffolding (Figure 6.8c). The release arrives at the plateau after ten days and no significant release of phosphate was observed until 21 days. For 3D-PPF-Amine-GO@BP scaffolds, the phosphate release was found to be significantly more regular as compare to 3D-PPF-Amine-BP based scaffolds materials in the case of DI water (Figure 6.8c). Then, but to a lesser degree, phosphate releases continued (Liu et al. 2019).

The inhibited oxidation phenomenon of BP nanolayers after wrapping and burying under a GO based nanolayer can result in this modification. The BP nanosheet covering the GO layer can reduce the exchanges of fluid and thus limit the oxidization of buried BP nanosheets by H_2O and O_2. In comparison with 3D PPF-Amine scaffoldings in the absence of coating process of 2D material, immunoscope imaging process of pre-osteoblast cells signified the elongated morphology of the three forms of functionalized scaffolds material (Figure 6.8d); on the other hand, SEM images demonstrated that cells formed robust filament when in contact with BP nanosheets (Figures 6.8d). The attraction process of constant release of phosphate $(PO_4)^{3-}$ species from the nanolayers of the BP may be due to this morphology, which includes numerical analysis of indicators of the cell form, including the cell circularity, cell length, and cell aspect ratio, which show that the cells have a longer line-shaped form in three types of functional scaffolding than the 3D-PPF-amine-based scaffolds materials in the absence of 2D material-based coating. Before observing the proliferation density values on the stated scaffolds-based material, MC_3T_3-E1 preosteoblast cells were discolored with anti-vinculin–FITC antibody and reacted with RP to label vinculin (red) and cellular filaments (green). Immunofluorescence-based reflection employing a confocal imaging process represented the clear cell propagation process after one, three, and six days time duration for culture growth on these scaffolds (Figure 6.8e). As shown in Figure 6.8e, cells were attached and distributed through scaffolds in all four-scaffold forms on both borders and cross-sections. Yet, there were very different cell numbers on these scaffolds' materials. For 3D-PPF-Amine-based scaffolds material in the absence of any 2D materials, the cells' proliferation process moved toward a higher density relative to scaffolds without 2D materials 3D-PPF-Amine-GO-based scaffolds material terminated with GO-based-only components. This is consistent with published research on carbon material-functionalized substrates that observed increased cell proliferation (Liu, Kim, et al. 2018). Cells on 3D-PPF-Amine-BP scaffolds have advanced toward the enhanced density value as compared with scaffolds material in the absence of 2D materials, close to the

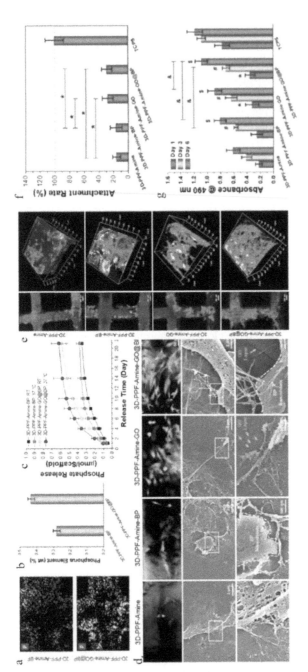

FIGURE 6.8 a) Phosphorus elemental mapping functionalized with BP and GO@BP nanosheets on 3D scaffolds and b) representing the estimation process of phosphorus content. c) Phosphate release kinetics at room temperature and body temperature 37°C (n = 2) from 3D scaffolds with BP and GO@BP nanosheets functionalized. d) Immunofluorescence and SEM images of MC_3T_3 pre-osteoblast cell morphology after one day of growth on functionalized 3D scaffolds indicate the edges of BP nanosheets; and GO layers e) Immunofluorescence and 3D reconstruction images of MC_3T_3 pre-osteoblasts on functionalized 3D scaffolds after six days of growth. f) At six hours post-seeding, cell attachment rate (n = 4 samples; *: p < 0.05). g) Proliferation of cells on functionalized 3D scaffolds at one, three, and six days post-seeding (n = 4 samples; *, #, $: p < 0.05 as equating to the 3D-PPF-Amine species at a time duration of days one, three, and six, respectively; &: p < 0.05). Adapted with permission from Liu et al. (2019), Copyright © 2019, American Chemical Society.

3D-PPF-Amine-GO-based scaffolds material. Cells proliferated process toward higher density value with confluent cells covering nearly all of the surface area for 3D-PPF-Amine-GO@BP scaffolds for both BP and GO nanosheets (Liu et al. 2019).

3D reconstruction-based imaging process using z-axis depth scanning revealed that all the scaffold's concerned boundaries were covered by cells. This outcome suggests that on the 3D scaffolds, GO and BP could synergistically increase cell manufacturing process. It also examined the adhesion process of MC_3T_3 pre-osteoblasts to 3D-based scaffolds material coated with 2D nanostructures. As illustrated in Figure 6.8f, there was substantially more cell adhesion in the 3D-PPF-Amine-GO@BP and 3D-PPF-Amine-GO scaffolds, all coated with a coating of GO. However, no increase in cell adhesion was detected for 3D scaffolds material coated with BP alone instead of 3D-PPF-Amine scaffolds material in the absence of 2D material-based coating, suggesting that GO works primarily to boost the initial process of cell adhesion. The one-day post-seeding quantitative cell number assessment on scaffolds verified that 3D-PPF-Amine-GO and 3D-PPFAmine-GO@BP scaffolds-based nanomaterial had meaningfully higher value of cell densities (Figure 6.8g). All the mentioned three categories of functionalized scaffolds displayed meaningfully greater importance of cell densities at three days of cultivation related to scaffolds material without the coating of 2D nanomaterial. Cell densities were also significantly higher on 3D-PPF-Amine-GO@BP scaffolds than on 3D-PPF-Amine-BP scaffolds (Figure 6.8g).

Compared with all other scaffolds, 3D-PPF-Amine-GO@BP scaffolds had the highest values of cell densities at six days duration of culture growth, i.e., experiments in which only one component of BP or GO has been functionalized or non-functionalized (Figure 6.8g). The above findings have enhanced the adsorption process of protein and cell adhesion by addition of GO nanosheets to the scaffold, taking the benefit of the greater surface area of the 2D layer of GO material (Liu et al. 2019).

On the other hand, by adding BP nanosheets, scaffolds were released as the concerned material is facing the process of degradation, causing a stretched cell structure and the formation of cellular filaments around their boundaries. Furthermore, a 3D confocal scan and cellular marker measurements provided by osteogenesis have shown that scaffolds functioning both with BP and GO nanosheets achieve the maximum robust cell relation, propagation, and differentiation. These findings have clearly shown that 2D GO and BP can synergistically stimulate cell growth and osteogenesis and make this an encouraging route for tissue engineering applications.

6.2.3   ENERGY STORAGE DEVICES

6.2.3.1   Li-Ion Batteries

The presence of diverse characteristics like high energy density, excellent cycling performance, and high operating voltage mean that lithium-ion batteries (LIBs) have been the subject of increasing attention due to their feasibility in potential applications. However, there are still some problems associated with lithium-ion batteries' performance, such as the low electrical conductivity caused by the vast volume expansion and repeated circulation. The presence of these problems reduces the practical applications of the fabricated device. Therefore, to get efficient results according to the market demand, developing some advanced type of anode materials associated with a stable life cycle and reliable specific capacity is necessary. To get the required target, various 2D materials were applied, and among them, BP has found efficient results in electronics-related fields like capacitors, transistors, sensors, and Li-S batteries. Although there are many advantages, there are still some issues that exist that need to be resolve in the case of LIBs (Xu et al. 2016; Liu, Hu, et al. 2018).

For example, large volumetric-based variations around 300 percent exist during the intercalation process of lithium-ion. Regarding the specific geometry of BP in the form of lamellar structure, there is a certain gap between the intercalation process related to li-ion between the layers (Sun et al. 2012). This process is responsible for the variation in volume. Subsequently, the active material is degraded and moves toward the surface layers after a long-term cycling process and cause the electrodes' cycling instability process. Different contributions have been implemented to overcome this problem, such as metal ion modification, electrode surface modification and nanocomposites recombination, etc. For example, Guo et al. (2017) published a report on the metal ion-based modification process to recover BP's characteristics. In another study, Xing et al. (2017) developed the BP-based nanocomposite for enhancing the thermal stability process. And in another approach, Wang et al. applied the reported p-Toluene sulfonyl isocyanate function as an electrolyte-based additive material to minimize the decomposition process of electrolytes as well as the cracking process of the solid electrolyte interface (SEI) (Wang, Li, et al. 2017). Furthermore, different varieties of the carbon-based nanomaterial are successfully engaged with BP, based on excellent electrical conductivity as well as mechanical properties (Wang, Jackman, et al. 2017; Ji et al. 2015). For instance Zhang et al. (2020) successfully combined carbon nanotubes (CNTs) and BP to develop a 3D form combined network. Electrostatic adsorption and cross-linking produce a stable chemical interaction of conductive network, performing a significant role in an electrically conductive network and 3D support framework. The computational (DFT) approach reveals the strong binding interaction of CNTs and BP. The phenomenon can enhance the electrical conductivity and reversible electron and ion transport phenomenon inside the active material layer interface to place and reduce the volume expansion and keep the BP structure safe during the lithiation and delithiation process. Figure 6.9a shows the charge transfer phenomenon of multilayer BP (left) and the hybrid structure of BP-CNTs (right). The long route and slow charge transfer phenomenon are developed in the case of original layers of BP represented on the left side; on the other hand, a brief and quick-track path flourished in the hybrid structure of BP@CNTs as shown on the right. Due to the improved value of electrical conductivity and shortened charge transfer track of CNTs, the electrical conductivity and charge transfer characteristics of the hybrid CNTS -BP improve the CV curves of separate CNTs, BP. The combined form of BP@CNTs for the first cycle results at a scanning speed of 0.2 mV s^{-1} with a range of working potential from 0.01 to 2.0 V are presented in Figure 6.9b. In contrast, the hybrid's energy storage process based on BP@CNTs is gradually enhanced due to enhancing the CV curve's redox-based reaction. It is observed that as a result of the discharge (Li insertion) and charge process (Li extraction) curves process, two reduction peaks (1.53 and 0.01 V) and two oxidation peaks (0.28 and 1.86 V) appear for CNTs. Furthermore, about four reduction

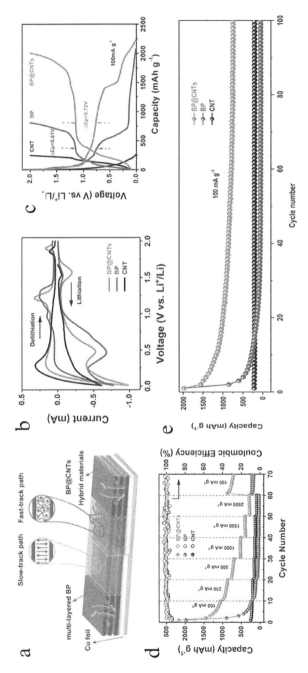

FIGURE 6.9 a) Electrochemical tests and characterization process of for individual structure of CNTs, BP, and hybrid sample BP@CNTs as LIB anodes. b) The cyclic voltammograms curve of CNTs, BP, and hybrid samples BP@CNTs. c) Galvanostatic-based charge/discharge cures for the initial cycle. d) Rate performance and coulombic efficiency at various current densities. e) Galvanostatic discharge profiles at a current density value of 100 mA g⁻¹ within a potential window range of 0.01–2 V (vs Li+/Li0) of individual structure of CNTs, BP, and hybrids electrodes network of BP@CNTs. Adapted with permission from Zhang et al. (2020), Copyright © 2020, WILEY–VCH Verlag GmbH & Co. KGaA, Weinheim.

peaks appeared at various potentials (1.53, 0.47, 0.22, and 0.01 V) and four oxidation peaks (0.31, 0.96, 1.23, and 1.90 V) for BP-based electrodes, follow the equivalent state: BP $\leftrightarrow$ LixP $\leftrightarrow$ LiP $\leftrightarrow$ Li$_2$P $\leftrightarrow$ Li$_3$P (Park and Sohn 2007).

Figure 6.9c represents the galvanostatic-based charge and discharge curves related to the separate CNTs, BP, and combined structure form of BP@CNTs. The value of specific capacity concerned with the hybrids network based on the BP@CNTs network is considered according to participants' weight contents, i.e., BP@CNTs. The value of the charge/discharge capacity of the CNTs-based electrode system at 100 mA g^{-1} is 260.7 and 247.7 mA h g^{-1} respectively. On the other hand, the BP-based electrodes network's BP charge/discharge curve value is similar to the BP nanomaterials. It is observed that as compared to other samples, the BP contents-associated CNTs (BP@CNTs) via P-C bonding has an initially noted value of discharge specific capacity of 2229 mAh g^{-1} at a potential range value of 0.01 and 2.0 V. The reversible charge specific capacity value is found around 2004 mAh g^{-1}, the initial coulombic efficiency (ICE) value (about 90 percent) is relatively more significant than the value of BP-based electrodes.

The BP@CNTs material shows a higher discharge capacity as compared to original CNTs and BP. The hybrid BP@CNTs electrode represents a high attained reliable electrical conductivity value (88.9 Ω), an excellent value of rate performance value (552 mAh g^{-1} at a current density value of 2.5 A g^{-1}), discharge capacitance value (1088 mAh g^{-1} at 0.1 A g^{-1}), long-term value of cycling stability value (757.3 mAh g^{-1} at 0.5 A g^{-1} for 650 cycles), and efficient energy density value (543 Wh kg^{-1} at 0.1 A g^{-1}) as well (Figure 6.9d–e) (Zhang et al. 2020).

The hybrid structure's efficient performance relates to good affinity, intense trapping, high adsorption of polyphosphoric, strong trapping, and intense interaction between the BP and CNTs. These findings provide a new direction of research relating to the preparation process of highly efficient electrodes, useful for optoelectronics and energy storage devices.

6.2.4 SUPERCAPACITORS

During the past decade, old-fashioned stretchable and bendable supercapacitors with smart and thin electrodes having a 2D shape are incompatible with 3D wearables due to limited specific areal capacitance (Huang et al. 2016). Lv, Tang, et al. (2018) developed a 3D stretchable supercapacitor made with polypyrrole/black-phosphorus oxide combined with carbon nanotube (CNT) film through the electrodeposited process, which was constructed on a stretchy honeycomb composite electrode and encouraged by a honeycomb lantern. The device-thickness-independent ion-transport path and stretchability of 3D stretchable supercapacitors can be molded into adjustable device thicknesses to improve precise areal energy storage and wearability.

The porous composite of PPy/BPO-CNT electrode results in 3D flexible and stretchable supercapacitor electrodes due to high flexibility, stress-relieving functions, and fast ion transport properties. The PPy/BPO-CNT composite conductor was collected into 3D flexible supercapacitors to maintain the important role of porous structure for the electrochemical and mechanical properties of flexible supercapacitors, as the rectangular shape of 3D stretchable supercapacitors with multilayer supercapacitor units as shown in Figure 6.10a,b (Lv, Tang, et al. 2018; Boota et al. 2016; Wang et al. 2015; Wang, Wu, et al. 2017; Lang et al. 2010). After that, when expansion occurs in the rectangular-shaped supercapacitor due to a honeycomb-like construction, hexagonal components are designed. Figure 6.10c illustrates the shape-independent/ device-thickness stretchability and confirmed it (after an increase in the thickness (T) of the device ranging from 0.1 to 3.0 cm) and found that thickness did not have any interaction with the rupture strain. Moreover, in the rectangle-shaped (having the thickness of 3 cm) supercapacitor, the maximum amount/quantity of the mass loading of a composite electrode might be reached to a range of 453 ± 36 mg cm^{-2}. There is a direct relation observed in Figure 6.10d ($R^2 = 0.9994$) between the device thickness and specific areal capacitance in the rectangular-shaped supercapacitors for conducted experimental results, and these findings suggest the possible 3D strategy (improving the specific areal capacitances) of stretchable supercapacitors. At different strains, rectangular-shaped supercapacitors composed of PPy/BPO-CNT electrodes have specific capacitances (normalized to 1), which were measured and have a comparison with the other two composite electrodes (PPy-based) such as PPy/SO$_4$$^{2-}$-CNT and PPy functionalized GO-CNT (as mentioned in Figure 6.10 b and 6.10e). It was found that without any electrochemical presentation, the degradation process of the supercapacitors established as a result of PPy functionalized BPO-CNT conductors was accomplished by being overextended to 2400 percent. Additionally, a large value of capacitance declines were found (decreased by 13 percent and 21 percent for PPy functionalized GO-CNT and PPy/SO4^{2-}-CNT based supercapacitors witnessed, respectively) (Cordeiro et al. 2015; Wood et al. 2014; Lv, Tang, et al. 2018). The major contribution of porous and composite PPy/BPO-CNT conductors might be due to the substantial drop in the efficiency of electrochemical performance. The expanded honeycomb structure paired with the highly flexible PPy/ BPO-CNT electrodes provides the 3D stretchable supercapacitors with a powerful integration of electrochemicals and mechanical durability.

The cycle stability in thicker rectangular supercapacitors is illustrated in Figure 6.10f, further shown by a repetition of 2,000 percent strain cycling tests (with geometric parameters of y = 0.7 cm, m = 0.2 cm, x = 194 μm, T = 1.0 cm).It was noted that even after increasing the number of 10,000 cycles as a result of the constant stretching test, around 1.0 cm thickness-size and rectangular-shape supercapacitors have the constant value of capacitance retention, which is around 95 percent. The value of capacitance in case of unscratched supercapacitors (97 percent) remains very close to that after 10,000 cycles indicating the uniform orientation of CNT and is responsible for the excellent electrochemical and mechanical stability of the PPy/ BPO-CNT electrode. Moreover, as a result of minor rise in the inside-created resistance value of the supercapacitor ranging 7.85 to 11.19 Ω there is an improvement in the inert presentation efficiency of the 3D stretchable and flexible supercapacitors. Rectangular-shaped supercapacitors with a fixed value thickness of 1.0 cm have the capacitance value of 7.35 F cm^{-2}, at a fixed value of current density around 7.8 mA/ cm2 have a conform value of the energy density value of 653 μWh/cm^2, at a constant value of power density 5.27 mW/cm^2. It is found that there is an increase in specific areal capacitance of 3D structure-based supercapacitors as compared to 2D-based

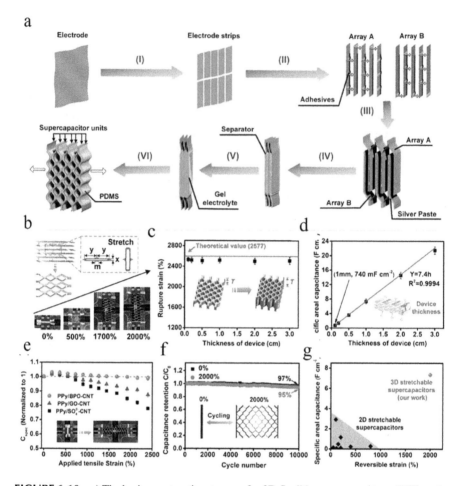

FIGURE 6.10 a) The basic construction process for 3D flexible supercapacitors: (I) First, the PPy-based electrodes were molded into identical electrode strips with predetermined shapes. (II) The electrode strips having a connection into two arrays system (electrode array A and electrode array B) and observed by adhesives (PVA/H$_3$PO$_4$ gel electrolyte). (III) In order to ensure the conductive connections, the edge length of both the arrays (A and B) are coated with paste of silver (Ag) ink, and after that, there is a crossed interaction between electrode array A and electrode array B. (IV) For the separation process of two arrays, a thin nano cellulose partition (about the length of 20 μm) is used. (V) Both the arrays (A and B) were laminated with PVA/H$_3$PO$_4$ gel electrolyte for developing a 3D stretchable based supercapacitor with bendable and expandable honeycomb-shaped structures. (VI) The 3D flexible supercapacitor was packed with a thin-layer PDMS. b) Under different strain tests the snapshot of supercapacitors being rectangular-shaped (having the geometric parameters of, m = 0.2 cm, y = 0.7 cm, x = 194 μm, T = 0.5 cm). The expandable honeycomb structure before and after being stretched, the hexagonal unit cell is shown in the inset images (upper left). c) On different places having various thicknesses of the device in the 3D stretchable based supercapacitors the requirement of rupture strain (having the geometric factors of, m = 0.2 cm, y = 0.7 cm, x = 194 μm). d) At a fixed current density value of 7.8 mA/cm², there is a performance-based assessment test of

FIGURE 6.10 Continued

3D stretchable based supercapacitors (having different values of device thicknesses). e) The normalized value of capacitance of different PPy functionalized electrodes with rectangular-shaped stretchable supercapacitors is confirmed at 7.8 mA/cm^2 under various value of strain condition. f) The retaining ratio of capacitance of PPy/BPO-CNT-based 3D stretchable supercapacitor (having geometric parameters of, m = 0.2 cm, x = 194 μm, y = 0.7 cm, T = 1.0 cm) is tested at fixed value of current density (7.8 mA cm) under the cycling tensile strain value of 2000 percent. g) As a determination of irregular stretchability value of 3D stretchable and flexible supercapacitor the specific areal capacitance has appeared (Lv, Luo, et al. 2018; Qi et al. 2015; Chen et al. 2013; Gilshteyn et al. 2016; Xiao et al. 2018; Tang et al. 2015; Meng et al. 2013; Choi et al. 2015; Lamberti et al. 2016; Choi et al. 2017; Lee et al. 2016; Kim et al. 2017; Qi et al. 2018; Lv, Tang, et al. 2018). Adapted with permission from Lv, Luo, et al. (2018) Copyright © 2018, WILEY–VCH Verlag GmbH & Co. KGaA, Weinheim.

supercapacitors and it is 61 times more than its original 2D (120 mF/cm^2 at the value of current density of 5 mA/cm^2) (Figure 6.10g).

A large number of supercapacitor units in both the parallel and series arrangements are another significant achievement attained due to stretchy honeycomb rectangle-shaped architecture for 3D stretchable supercapacitors, tuning the discharging time and output voltage to attain a position as per the requirement of related electronics (Lv, Tang, et al. 2018; Lv, Luo, et al. 2018; Qi et al. 2015; Chen et al. 2013; Gilshteyn et al. 2016; Xiao et al. 2018; Tang et al. 2015; Meng et al. 2013; Choi et al. 2015; Lamberti et al. 2016; Choi et al. 2017; Lee et al. 2016; Kim et al. 2017; Qi et al. 2018).

Figure 6.11a illustrated that there is a growth in voltage to 2.4 V, under the same value of charging/discharging current (there is a series connection of three super capacitor units) and capacitance of three super capacitor units (connected in stretched form to form a honeycomb stretchable structure) was observed. Consequently, the combination of parallel and in series connections increased discharging time and output voltage simultaneously by a factor of about three, as shown in Figure 6.11a. For confirmation, four supercapacitors (rectangular in shape) having geometric parameters of, m = 0.2 cm, y = 1.5 cm, x = 194 μm, T = 0.5 cm), were linked for a test to turn on a red light-emitting diode (LED) and it was analyzed that the supercapacitors (rectangular-shaped) stayed in process due to their high stretch-ability, concurrent double bending/stretching, and drive the LED with circulating Figure 6.11b. Unlike traditional stretchy supercapacitors that can only be formed in 2D planes, the newly created supercapacitors can be shaped into any 3D shape without losing their stretchability. Other 3D shapes were created using the expanded honeycomb structures in addition to the rectangular ones (Figure 6.11c–d). Besides, five stretchable supercapacitors with chair-shapes are mentioned, having a connection in a series outstretched like an accordion and can be bent as shown in Figure 6.11c. Moreover, the developed 3D-based flexible supercapacitor deforms into an arched, bridge-like shape (Figure 6.11d), and such shaped supercapacitors are associated with a ball-like geometry (arranged four units arranged in series). As a result, there is a process of deformation in supercapacitors (3D space created into a bowl-shaped con-struction, a volcano-shaped structure of balloon; Figure 6.11d) (Lv, Tang, et al. 2018).

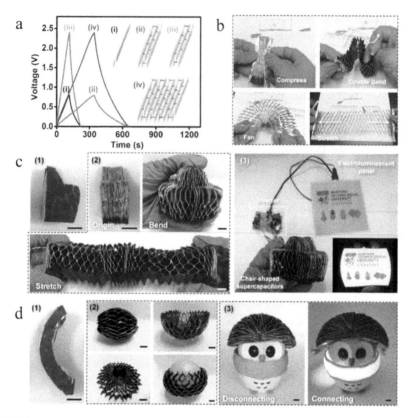

FIGURE 6.11 a) Galvanostatic charging/discharging curves (plotted between time and voltage) of (i) Curve represent the single unit supercapacitor, (ii)–(iv) when the supercapacitor units having a connection in series (3 × 1), in parallel (1 × 3) finally at a fixed value of current density (7.8 mA/ cm²) combined in-series/in parallel (3 × 3). b) With 48 supercapacitor units (4 × 12) the rectangle-shaped stretchable supercapacitors powering red LED under various stretching and bending states. c) Chair-shape 3D supercapacitors (stretchable): (1) The shape of the supercapacitor unit is taken from the optical microscope; (2) Under stretching and bending conditions the optical microscope images (chair like shaped stretchable and flexible supercapacitors) can be prolonged like an accordion shape (a portable box-like shape) morphology; (3) In order to supply power to an electrograph panel (10 cm × 10 cm), the supercapacitors (chair-shaped) are associated with an power inverter (power supply source) under bending state. d) Bridge-like curved supercapacitors: (1) The optical microscopic image (taken from optical microscope) of the of the unit of supercapacitor; (2) In upper left portion the supercapacitors are associated into a ball-like shape architecture (after that which can be breakdown further into a construction of bowl-like) as shown (upper right side), the surface of 3D balloon is confirmed from the volcano-like shape structure (down left) and (down right side); (3) The bridge-like supercapacitors (like a 3D helmet adjusted on the head of the bird toy-like model like an owl) and used to power a 3.0 V flexible and stretchable LED strip (right). Adapted with permission from Lv, Tang, et al. (2018), Copyright © 2018, WILEY–VCH Verlag GmbH & Co. KGaA, Weinheim.

Furthermore, using a power capacity of 3 V, the flexible and stretchable blue LED strip is arranged on the crown of a toy; these arched supercapacitors (bridge-like shape) can develop to function as like a helmet (Figure 6.11d).

In another study, Wen et al. (2019) introduced a simple electrochemical technique and synthesized 3D semi-associated BP sponges in a short time under an ambient atmosphere. The resultant sponge with controlled morphology with controllable size and shape (having a length over 10μm (large) and thickness is less than 4.0 nm (ultrathin) BP-based nanosheet-like structures) and there is a supply of semi-connected nanosheet-like constructions of open channels. In all the solid-state supercapacitor, the BP sponge has a specific capacitance value of 80 F/g at 10 mV/s as electrode material and the mentioned value is more sophisticated than those of 2D BP nanosheets and bulk BP. Good energy storage properties might be expected from 3D BP sponges, comprising ultrathin/large BP nanosheets and a large number of channels and, for confirmation and demonstration (Figure. 6.12a), an all-solid-state type supercapacitor based on BP based sponge conductor (BP-ASSP) is organized.

The BP-based sponge is separated into small pieces, dispersed in ethanol, and finally distributed on the PET substrate coated with gold. A polyvinyl alcohol/phosphoric acid (PVA/H_3PO_4) gel electrolyte is introduced to penetrate the BP sponge electrodes and for a sandwiched structure electrodes are pressed together for the fabrication of all-solid-state supercapacitor. Cyclic voltammetry (CV) used for estimating the efficiency of BP-ASSP is shown in Figure 6.12b. The CV bends show the characteristic rectangular electrochemical double-layer capacitance bends at different rates of scanning ranging between 10 and 100 mV/s. Figure. 6.12c shows the consistent mass capacitance value of BP-ASSP at the different values of scanning rates and the BP-ASSP has a specific value of capacitance 80 F g^1 at a fixed scanning rate of 10 mV s^1. Although there is an inverse relationship between the calculated capacitance and scanning rate, at 100 mV s^1, a relatively high capacitance value of around 28 F g^1 is detected. Figure 6.12d shows at the various current densities the values of the galvanostatic discharging/charging (GCD) curves of the BP-ASSP. The GCD curves during adsorption/desorption are good symbols for charge dispersion at the BP nanosheet/H_3PO_4 interfaces. The dispersion of BP-based all-solid-state supercapacitors (BP-ASSP) is considered at an extraordinary scanning rate value of 100 mV s^1 and the efficiency-specific value of capacitance retaining is around 80 percent after passing cycles around 15,000 charging/discharging. This evidence strongly confirms a better changeability (Figure 6.12e).

The factors that effect the performance/efficiency of 3D/2D BP sponge in case of all the solid-state supercapacitors are as follows:

1. High carrier mobility and large ratio of exposed atoms provide by the BP sponge's ultrathin nanosheets.
2. The size of BP nanosheet in the sponge is crucial for the interfacial electron transfer kinetics and the BP units in the sponge (having a large dimension range tens of micrometers) are found as suitable due to the referred extraordinary energy obstacles among the single nanosheets.
3. The process of promoting ion transport and electrolytic diffusion is made successful by the pores and channels in the BP sponge. Despite this, much lower

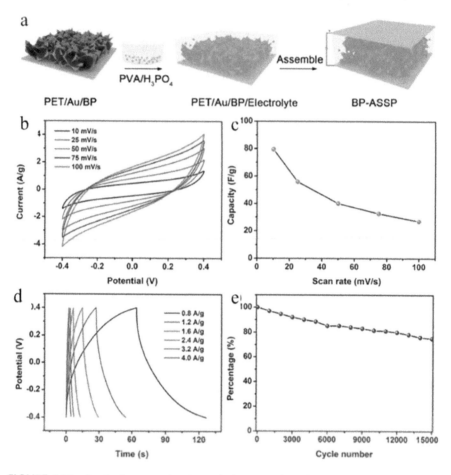

FIGURE 6.12 Synthetic protocol and practical presentation of the BP-ASSP. a) Schematic illustration of 3D BP-based sponges, comprising ultrathin/large BP nanosheets and a large number of channels. b) With different scanning rate cyclic voltammograms (CVs) curves were developed. c) The equivalent value of mass capacitances for BP-ASSP at the various scanning rates and the BP-ASSP has a specific value of capacitance (80 F g¹) at a fixed scanning rate of 10 mV s¹. d) The various current densities the galvanostatic charging/discharging (GCD) bends of the BP-ASSP. e) The BP-ASSP is considered at an extraordinary value of scanning rate (100 mV s¹) and the efficiency exact capacitance retaining is around 80 percent after 15,000 charging/discharging rounds. Adapted with permission from the Royal Society of Chemistry (Wen et al. 2019).

mass current densities were found in all-solid-state-based supercapacitors, including the bulk crystal of BP or 2D nanosheets of BP-like structure as the stretchable electrode-based materials. There is great potential in high-quality BP sponges for energy storage applications and it might be helpful for the improvement of BP-based technological applications.

In summary for future study, the established strategies of 3D stretchable supercapacitors could be used potentially for the construction process of other 3D stretchable energy storage device structures such as lithium-ion batteries (Song et al. 2015; Xu et al. 2013; Zhu, Tang, et al. 2018; Wei et al. 2018; Qian et al. 2018; Tang et al. 2016; Tang et al. 2018), and also might be incorporated with other flexible and stretchable devices like energy gleaning devices (Zhang et al. 2015; Guo, Yeh, et al. 2016) and wearable and chemical sensors.(Jiang et al. 2018; Liu, Qi, et al. 2018; Liu, Qi, et al. 2015; Pan et al. 2016). Other purposes—such as self-healing based shape memory, electrochromic and photo-/thermos-responsiveness—might be further integrated to build smarter 3D porous structures using the 3D stretchable supercapacitors for energy storage devices (Yang et al. 2017; Wang et al. 2014; Zhang et al. 2017).

6.2.5 CATALYSIS

6.2.5.1 Electrocatalysis

During the past decade, the designing of a high-performance palladium (Pd)-based catalyst used for enhancing the ethanol oxidation reaction (EOR) used in many industrial products was a big challenge. Wu et al. (2018) developed an innovative anatase titanium dioxide nanosheets-BP (ATN-BP)-based hybrid and used it as a maintenance for palladium nanoparticles (PdNPs) for the EOR. Different techniques were used to confirm the interaction of 3D ATN–BP during the oxidation process of ethanol. Cyclic voltammograms (CVs) were first introduced to examine the EOR efficiency of the pairs of Pd:C, BP:Pd, ATN:Pd, and ATN-BP:Pd compounds/catalyst in terms of mass activity and electrochemically active area (ECSA). Under the influence of N_2 gas, using the voltage in the range of (−0.1 and 0.2 V) in a 1.0 mol/L NaOH solution with a scanning rate of 50 mV/s the process of measurements was conducted. Figure 6.13a shows the CVs of the pair of catalysts such as (Pd:C, BP:Pd, Pd:ATN, and ATN-BP: Pd) in a couple of form of well-defined and visualized current crests labelled as Peaks I and II, between −1.0 and −0.7 V, which represented the hydrogen adsorption/desorption process, and the Peak IV created as a consequence of the reduction process of Pd oxide through the process of reduction sweep having a central position at around −0.32 V.

It is clearly observed that the Pd oxidation/reduction peaks of the pair of a catalyst such as Pd:C, Pd:ATN, and BP:Pd were substantially lower than Pd:ATN-BP-20%. It is important to note that the Pd oxide (PdO) reduction peaks of the pair of a catalyst such as a catalyst Pd:ATN) and Pd:ATN-BP-20% completely shift by ≈60 mV when there is a comparison with the catalyst Pd/C. It might be due to the interaction of Pd with the O_{2-} bulk vacancies or there is a process of charge transfer between the ATN and Pd and that might enhance the reduction of PdO. Subsequently, the catalyst Pd:ATN-BP displays the highest oxidative power.

Despite the outcome of the CV, the catalyst Pd:ATN-BP-20% at approximately ≈−0.15 V there appears an anodic-type wave measure (with a peak potential appearances (Peak III)) and is widely expected that the mentioned peak might be linked to the varying of the −OH in ATN-BP for high level of oxidation such as −COOH or CO_2. The CVs curve of the Pd:ATN-BP catalysts (having the various feed

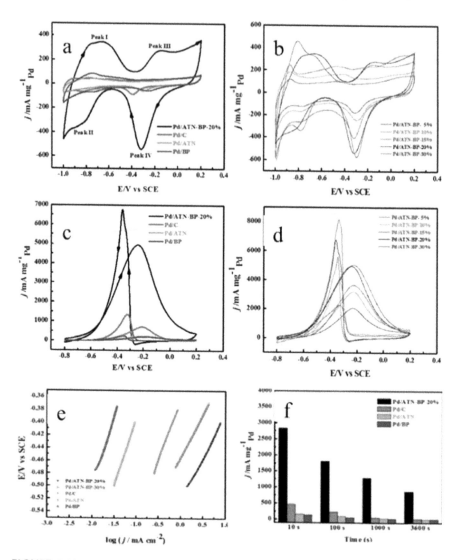

FIGURE 6.13 a) Cyclic voltammetry measurements (CVs) of different BP based catalysts such as pair of a catalyst such as Pd:C, Pd:BP, Pd:ATN, and Pd:ATN-BP in a basic1.0 M solution NaOH with a scan rate of 50 mV/s. b) CVs curves of Pd:ATN-BP catalysts in 1.0 mol/L basic solution NaOH at a scan rate of 50 mV/ s with changed feed mass percentages such as 5%, 10%, 15%, 20%, 30% of Pd. c) A combination of 1.0 M solution of NaOH combined with 1.0 mol/L EtOH solution at a scanning speed of 50 mV/s is used to investigate the electrocatalytic performance and efficiency of the ternary catalysts (Pd: ATN-BP) throughout the EOR process. The favorability of the electro-oxidation of ethanol is increased by the low potential value of catalyst Pd/ATN-BP-20%. d) The CVs measurement curves were acquired for Pd:ATN-BP-based catalysts as a result of feed mass percentages of Pd (5%, 10%, 15%, 20%, 30%) in combination of 1.0 mol/L solution of NaOH + 1.0 mol/L CH_3CH_2OH solution with a scan rate of 50 mV/s. e) At lower potentials of different catalysts derived from the CVs

FIGURE 6.13 Continued

of the pair of catalyst such as Pd:BP, Pd:ATN, Pd:C, and Pd:ATN-BP-20% (in a solution of 1 mol/L NaOH combined with 1 mol/LCH$_3$CH$_2$OH) shows a linear region value of the Tafel plots at a scanning rate of 10 mV/s (log j vs potential). f) There is assessment of the current density values of pairs of catalysts such as Pd:ATN, Pd:BP, Pd:C, and Pd:ATN-BP-20% for the EOR after robustness tests during the different time interval. Adapted with permission from Wu et al. (2018), Copyright © 2017, WILEY–VCH Verlag GmbH & Co. KGaA, Weinheim.

mass fractions of Pd) is shown in Figure 6.13b. As a result of obtained inductively coupled plasma-atomic emission spectrometry (ICP-AES) having various composition for the Pd, the loading contents of different catalysts (such as pairs of catalyst Pd:ATN, Pd:BP, Pd:ATN-BP-5%, Pd:ATN-BP-10%, Pd:ATN-BP-15%, Pd:ATN-BP-20%, and Pd:ATN-BP-30%) with different wt%, such as 9.6, 4.9, 4.6, 8.1, 10.2, 14.6, and 16.1 respectively were employed. It is clearly observed that after there is an increase in a loading content/amount of the Pd, the oxidation/reduction peaks and the areas of the hydrogen adsorption/desorption gradually increased. In addition, the important information concerning and related to the electrochemical active sites is provided by the electrochemically active area (ECSA). Comparatively, the ECSAPdO standards of Pd:ATN-BP-30%) (462.1) are much advanced than that of Pd:C (39.6), Pd:ATN (30.9), and Pd/BP (23.6 m2 gPd^{-1}). Additionally, the increase in the value of Pd:ATN-BP-30% (ECSA 462.1) might be due to the integrating of ATN-BP as an auxiliary material (which clearly has a significant consequence on the cumulative electrochemical activity of the catalyst). A combination of 1.0 M solution of NaOH with 1.0 mol/L solution EtOH solution at a scanning speed of 50 mV s^{-1} was used for investigating the electrocatalytic performance and efficiency of the ternary catalysts (Pd:ATN-BP) throughout the EOR process. The favorability of the electrooxidation of ethanol is increased by the low potential value of catalyst Pd:ATN-BP-20%, as shown in Figure 6.13c. As the beginning the potential of the pairs of catalyst such as Pd:ATN-BP-20% starts approximately at −0.57 V, the mentioned value is relatively lower than that of catalyst such as Pd:C (−0.49 V), Pd:ATN (−0.53 V), or Pd:BP (−0.55 V), and it might be due to the process of oxidation of the chemisorbed species resulting from ethanol. The process of oxidation of the chemisorbed species is confirmed due to the presence of an oxidation peak positioned at about −0.24 V in the forward direction of the scan and on the other side there is a reverse direction of scan peak related to the oxidation and desorption of carbonaceous (the species consisting of or yielding carbon) species.

In the case of Pd:ATN-BP-30% (5023.8 mA mg/ Pd) there is a proliferation in the value of mass peak current density assessment of the catalysts and it might be due to cumulatively loading the contents of Pd as shown in Figure 6.13d. Comparatively, the value of mass peak current density value of the catalyst Pd:ATN-BP-30% is almost 6.88, 20.59, and 24.62 times advanced than those of commercial pair of catalyst such as Pd:C (730.7 mA mg/Pd), Pd:ATN (244.03 mA mg/Pd), and Pd:BP (204.02 mA mg/Pd), respectively. Furthermore, due to these high values of current density, the resulting catalyst Pd/ATN-BP-30% is the most suitable candidate (among the mentioned pairs of catalyst) and showed reliable electrochemical activities for the

EOR. Furthermore, at lower potentials of different catalysts derived from the CVs of the pair of catalyst such as Pd:BP, Pd:ATN, Pd:C, and Pd:ATN-BP-20% (in a solution of 1 mol/L solution of NaOH combined with 1 M solution of CH_3CH_2OH) shows a linear region of the Tafel plots at a scan rate of 10 mV/s (log j vs potential) as shown in Figure 6.13e (Cai et al. 2015; Wu et al. 2018). Liang et al. (2009) proposed that the at lower potential values, the process of adsorption process of hydroxyl (–OH group) onto the conductors might be responsible for controlling the EOR kinetics, and further determined that the adsorbed –OH groups at the value of low potentials are responsible for the elimination of the adsorbed ethoxide anions species as well as its rate-defining step (Liang et al. 2009).

The Tafel slopes having the values of EOR on the above-mentioned pairs of catalysts such as Pd:ATN, Pd:BP, Pd:C, Pd:ATN-BP-20%, and Pd:ATN-BP-30% are noted as 205.8, 217.2, 187.6, 137.1, and 134.6 mV dec^{-1}, respectively. Fast charge-transfer kinetics is due to lower Tafel slope because of the rapid transmission of electron of the EOR and transport of active species on a catalyst (Shen, Zhao, and Xu 2010; Du et al. 2012; Jiang et al. 2016; Ren, Zhou, et al. 2017) and the lowest value of Tafel slope was found in the case of the Pd/ATN-BP-30% catalyst. This evidence strongly recommended that Pd/ATN-BP-30% has the fastest charge-transmission kinetics in the basic medium rather than among the Pd-based catalysts tested (Du et al. 2011; Du et al. 2014; Du et al. 2012).

Chronoamperometric curves were obtained in a combination of 1 M solution of NaOH and 1M solution of EtOH at a constant value of potential 0.2 V versus for 3600 s for the evaluation of the strength and stability of the catalyst Pd: ATN-BP. Moreover, the catalyst Pd: ATN-BP-20% demonstrated higher initial and final values of current densities values than the supplementary Pd-based catalyst, as shown in Figure 6.13f. This evidence is responsible for the highest electrochemical activity of the Pd:ATN-BP-based catalyst in the process of EOR.

It was analyzed that the EOR current density value in case of all the Pd-based catalysts is reduced suddenly within the first five minutes of scanning, and this is possibly due to the formation of Pd oxides (Wu et al. 2018). Finally, it was analyzed that for finding the highly efficient EOR catalysts, the combination of ATN-BP and PdNPs seems to be a very auspicious and attractive approach and further signifies that the development of appropriate heterointerfaces for the noble metal (Ag, Au, Pd, Pt) based catalysts can improve their electrochemical-based activity.

6.2.5.2 Photocatalysis

The presence of emerging characteristics in BP, such as the reliable value of charge mobility ≈ 1000 cm^2 V^{-1} s^{-1}, tunable bandgap, and remarkable Vis-NIR absorption capacity, makes BP an outstanding candidate used for photocatalysis (Liu, Du, et al. 2015; Li et al. 2014). (Zhu, Kim, et al. (2017) advance the 2DBP/CN-based heterostructure material for the photocatalytic-based hydrogen production process. In another approach, Boppella et al. (2019) also introduced the 2D hybrid $Ni_2P@BP@CN$ for the photocatalytic-based H_2 production. During the past decade, BP has been presented as an emerging 2D semiconducting material representing the reliable value of charge carrier mobility and wide tunable bandgap for photocatalysis. Although BP loses its stability and is degraded easily under the influence of air and water, it provides a suitable environment for the separation process of the photo-induced hole-electron pairs (Zhu, Zhang, et al. 2018).

Song et al. (2019) developed a facile approach to prepare the reticulated carbon nitride materials (CN-4N) for photocatalysis. The process of the self-capturing characteristics of CN-4N is utilized to capture the BP quantum dot (BQ) from the aqueous media. After that, a unique structure of BP is formed after the successful implantation process in the interior surface site of CN-4N by using the sonication process. After the successful sonication process, a 3D–0D organic-inorganic based hybrid structure was developed. The hybrid structure based on 3D/0D hybrid (CN-4N(BQ)) structure is found to be a reliable candidate for photocatalytic hydrogen production as equated to separate bulk graphitic carbon nitride materials (NCN) and CN-4N. The necessary steps involved during the preparation of the hybrid structure are shown in Figure 6.14a (Song et al. 2019).

It is observed that the multiple reelection process of light increased the ability of CN-4N porous structures to capture the light in the range of 450–550 nm as compared to the bare NCN structure. The CN-4N(BQ) represented the outstanding optical absorption potential in the range of Vis-NIR equated to the bulk CN-4N, facilitating the possible absorption phenomenon of low-energy vis-NIR. The excellent vis-NIR absorption potential of BQ has widely explained the exceptional absorption capacity of CN-4N(BQ). Figure 6.14b displays the NCN, CN-air, CN-4Ar, and CN-4N photo-catalytic H_2 development time courses. The mean H_2 evolution rate phenomenon of NCN was 0.39 mmol h^{-1} g^{-1}, and for CN-4Ar, CN-air, and CN-4N the value of evolution rate is 3.34 mmol h^{-1}g^{-1}, 1.20 mmol h^{-1} g^{-1} and 4.21 mmol h^{-1} g^{-1}, respectively. The value of the average rate of development of CN-4N for H_2 was 10.8-fold more significant than that of NCN. As illustrated in Figure 6.14c, for CN-4N(-4BQ), the mean manufacture rate of H_2 is enhanced to 8.06 mmol h1 g1 and the production action of H_2 is denoted by an upward tendency as the quantity of BQ increased (Song et al. 2019).

The attained CN-4N(BQ) demonstrated the best material for the photocatalytic based H_2 evolution behavior (13.83 mmol h^{-1} g^{-1}) once the BQ quantity reached 1.2 weight %, which is 3.3 times greater than that of bare CN-4N. Unfortunately, a steady rise in BQ quantity has contributed to a reverse trend in photocatalytic-based action, likely due to extreme recombination centers supplied by additional BQ. Remarkably, the average H_2 output rate value of CN-4N(BQ) was determined under the influence of NIR light illumination process (l > 700 nm) to be 0.47 mmol g^{-1} h^{-1}. Nevertheless, both the NCN and CN-4N are unable to display any photocatalytic-based activity due to the lack of absorption at wavelengths over 700 nm in the near-infrared range. Impressively, the same synthetic method used to acquire the CN-4N(BQ)(E) and CN-4N(BQ), but it was substituted with absolute ethanol, but the photocatalyst behavior was very different (Figure 6.14d). The CN-4N(BQ) showed apparent quantity efficiency (EQE) of 16.5 percent at 420 nm. The testing process related to the photocatalytic efficiency of CN-4N(BQ) was repeated for seven cycles to determine its stability and reusability, and there is no actual reduction process detected in the production rate of H_2, confirming the characteristics of the high stability phenomenon of CN-4N(BQ) (Figure 6.14e) (Song et al. 2019)

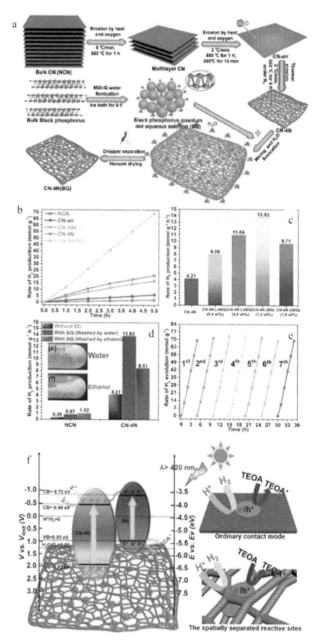

FIGURE 6.14 a) Schematic representation shows the preparation process of the composites of CN-4N(BQ). b) Courses of reaction time over the different aqueous solution-based photocatalysts. c) Rate of hydrogen evolution process on CN-4N filled with various contents of BQ. d) The photocatalytic activities of the samples NCN/BQ (1.2 wt percent BQ) and CN-4N(BQ) were washed separately with absolute ethanol and deionized water for the process of H$_2$ production. e) Cycling test of photocatalytic based H$_2$ production process for the CN-4N(BQ)-based hybrids material (Song et al. 2019).

According to the discussion, as mentioned earlier, a potential mechanism for photocatalytic H2 development under l > 420 nm irradiation was tentatively suggested (Figure 6.14 f). It is understood that both the internal surface of the CN-4N and the BQ's top surface could create photo-generated electro-holes. A reference standard equals 0 V versus reversible based hydrogen electrode (RHE) −4.44 eV versus vacuum level 1.65 with a radiation level greater than 420 nm. The photo-excited electrons were then transported to the conduction band (CB) of CN-4N rapidly on the BQ's top surface, since the CB of BQ was less negative than CN-4N. To achieve a photo-catalytic H_2 evolution reaction, the accumulated electrons eventually moved from the inner surface site of CN-4N to the outside surface. The results showed that BQ was magnificently captured on the CN-4N interior surface. In all the tested samples, in the case of CN-4N(BQ), the highest (13.83 mmol h^{-1} g^{-1}) H_2 manufacturing rate was acquired b, which was 3.3- and 35.5-fold advanced level for both the CN-4N and NCN, respectively (Song et al. 2019).

The plentiful active surfaces provided by the ultra-porous systems were necessary for the great H_2 evolution rate on the hybrids CN-4N(BQ) with their efficient Vis-NIR-based absorption process spatially segregating the reactive sites for the redox-based reaction and a highly improved photo-induced electron-hollow separation efficiency. This work offers a capable strategy for further exploiting and adapting CN-based hybrids for battery, electro-catalytic, solar, and desalination applications.

6.3   CONCLUSION

This chapter has provided a brief overview of optoelectronics, energy storage devices, biomedical applications, and catalysis-related applications of BP based hetrostructures with polymers, quantum dots, and metal oxides. Composites and hybrids in the form of 3D structures of BP with improved mechanical strength, electrical or thermal conductivity, light absorption or emission, or large-scale processing are produced by BP material filled or coated with matrix-like polymers, quantum dots, or metal oxides. Flexible and transparent substrates with limitless sizes and long lifetimes, as well as lower costs and easy planning, are expected to be the future of BP-based composite materials in photonic-related technology (potential and biomedical applications). As a result, the application of BP-based 3D structures is expected to remain a hot topic for the foreseeable future.

6.4   ACKNOWLEDGMENTS

The research was partially supported by financial support from the Science and Technology Development Fund (Nos. 007/2017/A1 and 132/2017/A3), Macao Special Administration Region (SAR), China, and National Natural Science Fund (Grant Nos. 61875138, 61435010, 61805147, and 6181101252), and the Science and Technology Innovation Commission of Shenzhen (KQTD2015 0324162703, JCYJ20150625103619275, JCYJ20180305125141661, and JCYJ20170811093453105). The authors also acknowledge the support from the Instrumental Analysis Center of Shenzhen University (Xili Campus).

REFERENCES

Boota, Muhammad, Babak Anasori, Cooper Voigt, Meng-Qiang Zhao, Michel W. Barsoum, and Yury Gogotsi. 2016. "Pseudocapacitive electrodes produced by oxidant-free polymerization of pyrrole between the layers of 2D titanium carbide (MXene)." *Advanced Materials* 28(7): 1517–1522.

Boppella, Ramireddy, Wooseok Yang, Jeiwan Tan, Hyeok-Chan Kwon, Jaemin Park, and Jooho Moon. 2019. "Black phosphorus supported Ni2P co-catalyst on graphitic carbon nitride enabling simultaneous boosting charge separation and surface reaction." *Applied Catalysis B: Environmental* 242: 422–430.

Cai, Bin, Dan Wen, Wei Liu, Anne-Kristin Herrmann, Albrecht Benad, and Alexander Eychmüller. 2015. "Function-led design of aerogels: self-assembly of alloyed pdni hollow nanospheres for efficient electrocatalysis." *Angewandte Chemie International Edition* 54(44): 13101–13105.

Castellanos-Gomez, Andres, Michele Buscema, Rianda Molenaar, Vibhor Singh, Laurens Janssen, Herre S. J. Van Der Zant, and Gary A. Steele. 2014. "Deterministic transfer of two-dimensional materials by all-dry viscoelastic stamping." *2D Materials* 1(1): 011002.

Chen, Che, Nathan Youngblood, Ruoming Peng, Daehan Yoo, Daniel A. Mohr, Timothy W. Johnson, Sang-Hyun Oh, and Mo Li. 2017. "Three-dimensional integration of black phosphorus photodetector with silicon photonics and nanoplasmonics." *Nano Letters* 17(2): 985–991.

Chen, Tao, Yuhua Xue, Ajit K. Roy, and Liming Dai. 2013. "Transparent and stretchable high-performance supercapacitors based on wrinkled graphene electrodes." *ACS Nano* 8(1): 1039–1046.

Chen, Yu, Chaoliang Tan, Hua Zhang, and Lianzhou Wang. 2015. "Two-dimensional graphene analogues for biomedical applications." *Chemical Society Reviews* 44(9): 2681–2701.

Choi, Changsoon, Ji Hwan Kim, Hyun Jun Sim, Jiangtao Di, Ray H. Baughman, and Seon Jeong Kim. 2017. "Microscopically buckled and macroscopically coiled fibers for ultrastretchable supercapacitors." *Advanced Energy Materials* 7(6): 1602021.

Choi, Changsoon, Shi Hyeong Kim, Hyeon Jun Sim, Jae Ah Lee, A. Young Choi, Youn Tae Kim, Xavier Lepró, Geoffrey M. Spinks, Ray H. Baughman, and Seon Jeong Kim. 2015. "Stretchable, weavable coiled carbon nanotube/MnO 2/polymer fiber solid-state supercapacitors." *Scientific Reports* 5: 9387.

Chung, Tze-Wen, Der-Zen Liu, Sin-Ya Wang, and Shoei-Shen Wang. 2003. "Enhancement of the growth of human endothelial cells by surface roughness at nanometer scale." *Biomaterials* 24(25): 4655–4661.

Cordeiro, Erlon R., Antonio W. C. Fernandes, Alessandra F. C. Pereira, Mateus M. da Costa, Marcio L. F. Nascimento, and Helinando P. de Oliveira. 2015. "Staphylococcus aureus biofilm formation on polypyrrole: an electrical overview." *Química Nova* 38(8): 1075–1079.

Das, Saptarshi, Wei Zhang, Marcel Demarteau, Axel Hoffmann, Madan Dubey, and Andreas Roelofs. 2014. "Tunable transport gap in phosphorene." *Nano Letters* 14(10): 5733–5739.

Deligianni, Despina D., Nikoleta D. Katsala, Petros G. Koutsoukos, and Yiannis F. Missirlis. 2000. "Effect of surface roughness of hydroxyapatite on human bone marrow cell adhesion, proliferation, differentiation and detachment strength." *Biomaterials* 22(1): 87–96.

Du, Wenxin, Qi Wang, David Saxner, N. Aaron Deskins, Dong Su, James E. Krzanowski, Anatoly I. Frenkel, and Xiaowei Teng. 2011. "Highly active iridium/iridium–tin/tin oxide heterogeneous nanoparticles as alternative electrocatalysts for the ethanol oxidation reaction." *Journal of the American Chemical Society* 133(38): 15172–15183.

Du, Wenxin, Kayla E. Mackenzie, Daniel F. Milano, N. Aaron Deskins, Dong Su, and Xiaowei Teng. 2012. "Palladium–tin alloyed catalysts for the ethanol oxidation reaction in an alkaline medium." *ACS Catalysis* 2(2): 287–297.

Du, Wenxin, Guangxing Yang, Emily Wong, N. Aaron Deskins, Anatoly I. Frenkel, Dong Su, and Xiaowei Teng. 2014. "Platinum-tin oxide core–shell catalysts for efficient electro-oxidation of ethanol." *Journal of the American Chemical Society* 136(31): 10862–10865.

Engel, Michael, Mathias Steiner, and Phaedon Avouris. 2014. "Black phosphorus photodetector for multispectral, high-resolution imaging." *Nano Letters* 14(11): 6414–6417.

Fiori, Gianluca, Francesco Bonaccorso, Giuseppe Iannaccone, Tomás Palacios, Daniel Neumaier, Alan Seabaugh, Sanjay K. Banerjee, and Luigi Colombo. 2014. "Electronics based on two-dimensional materials." *Nature Nanotechnology* 9(10): 768.

Freitag, Marcus, Tony Low, Fengnian Xia, and Phaedon Avouris. 2013. "Photoconductivity of biased graphene." *Nature Photonics* 7(1): 53.

Gilshteyn, Evgenia P., Tanja Kallio, Petri Kanninen, Ekaterina O. Fedorovskaya, Anton S. Anisimov, and Albert G. Nasibulin. 2016. "Stretchable and transparent supercapacitors based on aerosol synthesized single-walled carbon nanotube films." *RSC Advances* 6(96): 93915–93921.

Guo, Hengyu, Min-Hsin Yeh, Ying-Chih Lai, Yunlong Zi, Changsheng Wu, Zhen Wen, Chenguo Hu, and Zhong Lin Wang. 2016. "All-in-one shape-adaptive self-charging power package for wearable electronics." *ACS Nano* 10(11): 10580–10588.

Guo, Qiushi, Andreas Pospischil, Maruf Bhuiyan, Hao Jiang, He Tian, Damon Farmer, Bingchen Deng, Cheng Li, Shu-Jen Han, Han Wang, Qiangfei Xia, Tso-Ping Ma, Thomas Mueller, and Fengnian Xia. 2016. "Black phosphorus mid-infrared photodetectors with high gain." *Nano Letters* 16(7): 4648–4655.

Guo, Zhinan, Si Chen, Zhongzheng Wang, Zhenyu Yang, Fei Liu, Yanhua Xu, Jiahong Wang, Ya Yi, Han Zhang, and Lei Liao. 2017. "Metal-ion-modified black phosphorus with enhanced stability and transistor performance." *Advanced Materials* 29(42): 1703811.

Hao, Chunxue, Bingchao Yang, Fusheng Wen, Jianyong Xiang, Lei Li, Wenhong Wang, Zhongming Zeng, Bo Xu, Zhisheng Zhao, and Zhongyuan Liu. 2016. "Flexible all-solid-state supercapacitors based on liquid-exfoliated black-phosphorus nanoflakes." *Advanced Materials* 28(16): 3194–3201.

Huang, Xiao-Wei, Jing-Jing Wei, Meng-Ya Zhang, Xue-Liang Zhang, Xiao-Fei Yin, Chun-Hua Lu, Ji-Bin Song, Shu-Meng Bai, and Huang-Hao Yang. 2018. "Water-based black phosphorus hybrid nanosheets as a moldable platform for wound healing applications." *ACS Applied Materials & Interfaces* 10(41): 35495–35502.

Huang, Zhaodong, Hongshuai Hou, Yan Zhang, Chao Wang, Xiaoqing Qiu, and Xiaobo Ji. 2017. "Layer-tunable phosphorene modulated by the cation insertion rate as a sodium-storage anode." *Advanced Materials* 29(34): 1702372.

Huang, Zongyu, Zhen Zhang, Xiang Qi, Xiaohui Ren, Guanghua Xu, Pengbo Wan, Xiaoming Sun, and Han Zhang. 2016. "Wall-like hierarchical metal oxide nanosheet arrays grown on carbon cloth for excellent supercapacitor electrodes." *Nanoscale* 8(27): 13273–13279.

Ji, Yuanchun, Lujiang Huang, Jun Hu, Carsten Streb, and Yu-Fei Song. 2015. "Polyoxometalate-functionalized nanocarbon materials for energy conversion, energy storage and sensor systems." *Energy & Environmental Science* 8(3): 776–789.

Jiang, Qianqian, Lei Xu, Ning Chen, Han Zhang, Liming Dai, and Shuangyin Wang. 2016. "Facile synthesis of black phosphorus: an efficient electrocatalyst for the oxygen evolving reaction." *Angewandte Chemie* 128(44): 14053–14057.

Jiang, Ying, Zhiyuan Liu, Naoji Matsuhisa, Dianpeng Qi, Wan Ru Leow, Hui Yang, Jiancan Yu, Geng Chen, Yaqing Liu, and Changjin Wan. 2018. "Auxetic mechanical metamaterials to enhance sensitivity of stretchable strain sensors." *Advanced Materials* 30(12): 1706589.

Kim, Byoung Soo, Kangsuk Lee, Seulki Kang, Soyeon Lee, Jun Beom Pyo, In Suk Choi, Kookheon Char, Jong Hyuk Park, Sang-Soo Lee, and Jonghwi Lee. 2017. "2D reentrant auxetic structures of graphene/CNT networks for omnidirectionally stretchable supercapacitors." *Nanoscale* 9(35): 13272–13280.

Ku, Sook Hee, and Chan Beum Park. 2013. "Myoblast differentiation on graphene oxide." *Biomaterials* 34(8): 2017–2023.

Lamberti, Andrea, Francesca Clerici, Marco Fontana, and Luciano Scaltrito. 2016. "A highly stretchable supercapacitor using laser-induced graphene electrodes onto elastomeric substrate." *Advanced Energy Materials* 6(10): 1600050.

Lang, Xuemei, Qunyi Wan, Chunhua Feng, Xianjun Yue, Wendong Xu, Jing Li, and Shuanshi Fan. 2010. "The role of anthraquinone sulfonate dopants in promoting performance of polypyrrole composites as pseudo-capacitive electrode materials." *Synthetic Metals* 160(15–16): 1800–1804.

Lee, Habeom, Sukjoon Hong, Jinhwan Lee, Young Duk Suh, Jinhyeong Kwon, Hyunjin Moon, Hyeonseok Kim, Junyeob Yeo, and Seung Hwan Ko. 2016. "Highly stretchable and transparent supercapacitor by Ag–Au core–shell nanowire network with high electrochemical stability." *ACS Applied Materials & Interfaces* 8(24): 15449–15458.

Li, Hui, Kaat Fierens, Zhiyue Zhang, Nane Vanparijs, Martijn J. Schuijs, Katleen Van Steendam, Natàlia Feiner Gracia, Riet De Rycke, Thomas De Beer, and Ans De Beuckelaer. 2016. "Spontaneous protein adsorption on graphene oxide nanosheets allowing efficient intracellular vaccine protein delivery." *ACS Applied Materials & Interfaces* 8(2): 1147–1155.

Li, Likai, Yijun Yu, Guo Jun Ye, Qingqin Ge, Xuedong Ou, Hua Wu, Donglai Feng, Xian Hui Chen, and Yuanbo Zhang. 2014. "Black phosphorus field-effect transistors." *Nature Nanotechnology* 9: 372–377.

Li, Yulin, and Changsheng Liu. 2017. "Nanomaterial-based bone regeneration." *Nanoscale* 9(15): 4862–4874.

Liang, Z. X., T. S. Zhao, J. B. Xu, and L. D. Zhu. 2009. "Mechanism study of the ethanol oxidation reaction on palladium in alkaline media." *Electrochimica Acta* 54(8): 2203–2208.

Linez-Bataillon, P., F. Monchau, M. Bigerelle, and H. F. Hildebrand. 2002. "In vitro MC3T3 osteoblast adhesion with respect to surface roughness of Ti6Al4V substrates." *Biomolecular Engineering* 19(2–6): 133–141.

Liu, Han, Yuchen Du, Yexin Deng, and Peide D. Ye. 2015. "Semiconducting black phosphorus: synthesis, transport properties and electronic applications." *Chemical Society Reviews* 44(9): 2732–2743.

Liu, Hanwen, Kui Hu, Dafeng Yan, Ru Chen, Yuqin Zou, Hongbo Liu, and Shuangyin Wang. 2018. "Recent advances on black phosphorus for energy storage, catalysis, and sensor applications." *Advanced Materials* 30(32): 1800295.

Liu, Xifeng, Joseph C. Kim, A. Lee Miller, Brian E. Waletzki, and Lichun Lu. 2018. "Electrically conductive nanocomposite hydrogels embedded with functionalized carbon nanotubes for spinal cord injury." *New Journal of Chemistry* 42(21): 17671–17681.

Liu, Xifeng, A. Lee Miller, Sungjo Park, Matthew N. George, Brian E. Waletzki, Haocheng Xu, Andre Terzic, and Lichun Lu. 2019. "Two-dimensional black phosphorus and graphene oxide nanosheets synergistically enhance cell proliferation and osteogenesis on 3D printed scaffolds." *ACS Applied Materials & Interfaces* 11(26): 23558–23572.

Liu, Zhiyuan, Dianpeng Qi, Peizhi Guo, Yan Liu, Bowen Zhu, Hui Yang, Yaqing Liu, Bin Li, Chenguang Zhang, and Jiancan Yu. 2015. "Thickness-gradient films for high gauge factor stretchable strain sensors." *Advanced Materials* 27(40): 6230–6237.

Liu, Zhiyuan, Dianpeng Qi, Wan Ru Leow, Jiancan Yu, Michele Xiloyannnis, Leonardo Cappello, Yaqing Liu, Bowen Zhu, Ying Jiang, and Geng Chen. 2018. "3D-structured stretchable strain sensors for out-of-plane force detection." *Advanced Materials* 30(26): 1707285.

Luo, Miaomiao, Taojian Fan, Yun Zhou, Han Zhang, and Lin Mei. 2019. "2D black phosphorus–based biomedical applications." *Advanced Functional Materials* 29(13): 1808306.

Lv, Ruichan, Dan Yang, Piaoping Yang, Jiating Xu, Fei He, Shili Gai, Chunxia Li, Yunlu Dai, Guixin Yang, and Jun Lin. 2016. "Integration of upconversion nanoparticles and ultrathin black phosphorus for efficient photodynamic theranostics under 808 nm near-infrared light irradiation." *Chemistry of Materials* 28(13): 4724–4734.

Lv, Zhisheng, Yifei Luo, Yuxin Tang, Jiaqi Wei, Zhiqiang Zhu, Xinran Zhou, Wenlong Li, Yi Zeng, Wei Zhang, and Yanyan Zhang. 2018. "Editable supercapacitors with customizable stretchability based on mechanically strengthened ultralong MnO2 nanowire composite." *Advanced Materials* 30(2): 1704531.

Lv, Zhisheng, Yuxin Tang, Zhiqiang Zhu, Jiaqi Wei, Wenlong Li, Huarong Xia, Ying Jiang, Zhiyuan Liu, Yifei Luo, and Xiang Ge. 2018. "Honeycomb-lantern-inspired 3D stretchable supercapacitors with enhanced specific areal capacitance." *Advanced Materials* 30(50): 1805468.

Melikyan, Argishti, Luca Alloatti, Alban Muslija, David Hillerkuss, Philipp C. Schindler, J. Li, Robert Palmer, Dietmar Korn, Sascha Muehlbrandt, and Dries Van Thourhout. 2014. "High-speed plasmonic phase modulators." *Nature Photonics* 8(3): 229.

Meng, Yuning, Yang Zhao, Chuangang Hu, Huhu Cheng, Yue Hu, Zhipan Zhang, Gaoquan Shi, and Liangti Qu. 2013. "All-graphene core-sheath microfibers for all-solid-state, stretchable fibriform supercapacitors and wearable electronic textiles." *Advanced Materials* 25(16): 2326–2331.

Nam, Jutaek, Sejin Son, Lukasz J. Ochyl, Rui Kuai, Anna Schwendeman, and James J. Moon. 2018. "Chemo-photothermal therapy combination elicits anti-tumor immunity against advanced metastatic cancer." *Nature Communications* 9(1): 1–13.

Novoselov, Kostya S., D. Jiang, F. Schedin, T. J. Booth, V. V. Khotkevich, S. V. Morozov, and Andre K. Geim. 2005. "Two-dimensional atomic crystals." *Proceedings of the National Academy of Sciences* 102(30): 10451–10453.

Ou, Wenquan, Jeong Hoon Byeon, Raj Kumar Thapa, Sae Kwang Ku, Chul Soon Yong, and Jong Oh Kim. 2018. "Plug-and-play nanorization of coarse black phosphorus for targeted chemo-photoimmunotherapy of colorectal cancer." *ACS Nano* 12(10): 10061–10074.

Pan, Shaowu, Jing Ren, Xin Fang, and Huisheng Peng. 2016. "Integration: an effective strategy to develop multifunctional energy storage devices." *Advanced Energy Materials* 6(4): 1501867.

Park, C-M, and H-J Sohn. 2007. "Black phosphorus and its composite for lithium rechargeable batteries." *Advanced Materials* 19(18): 2465–2468.

Qi, Dianpeng, Zhiyuan Liu, Yan Liu, Wan Ru Leow, Bowen Zhu, Hui Yang, Jiancan Yu, Wei Wang, Hua Wang, and Shengyan Yin. 2015. "Suspended wavy graphene microribbons for highly stretchable microsupercapacitors." *Advanced Materials* 27(37): 5559–5566.

Qi, Ruijie, Jinhui Nie, Mingyang Liu, Mengyang Xia, and Xianmao Lu. 2018. "Stretchable V 2 O 5/PEDOT supercapacitors: a modular fabrication process and charging with triboelectric nanogenerators." *Nanoscale* 10(16): 7719–7725.

Qian, Guoyu, Bin Zhu, Xiangbiao Liao, Haowei Zhai, Arvind Srinivasan, Nathan Joseph Fritz, Qian Cheng, Mingqiang Ning, Boyu Qie, and Yi Li. 2018. "Bioinspired, spine-like, flexible, rechargeable lithium-ion batteries with high energy density." *Advanced Materials* 30(12): 1704947.

Qiao, Jingsi, Xianghua Kong, Zhi-Xin Hu, Feng Yang, and Wei Ji. 2014. "High-mobility transport anisotropy and linear dichroism in few-layer black phosphorus." *Nature Communications* 5: 4475.

Qiu, Meng, Dou Wang, Weiyuan Liang, Liping Liu, Yin Zhang, Xing Chen, David Kipkemoi Sang, Chenyang Xing, Zhongjun Li, and Biqin Dong. 2018. "Novel concept of the smart NIR-light–controlled drug release of black phosphorus nanostructure for cancer therapy." *Proceedings of the National Academy of Sciences* 115(3): 501–506.

Ren, Xiaohui, Zhongjun Li, Zongyu Huang, David Sang, Hui Qiao, Xiang Qi, Jianqing Li, Jianxin Zhong, and Han Zhang. 2017. "Environmentally robust black phosphorus nanosheets in solution: application for self-powered photodetector." *Advanced Functional Materials* 27(18): 1606834.

Ren, Xiaohui, Jie Zhou, Xiang Qi, Yundan Liu, Zongyu Huang, Zhongjun Li, Yanqi Ge, Sathish Chander Dhanabalan, Joice Sophia Ponraj, and Shuangyin Wang. 2017. "Few-layer black phosphorus nanosheets as electrocatalysts for highly efficient oxygen evolution reaction." *Advanced Energy Materials* 7(19): 1700396.

Sajedi-Moghaddam, Ali, Carmen C. Mayorga-Martinez, Zdeněk Sofer, Daniel Bouša, Esmaiel Saievar-Iranizad, and Martin Pumera. 2017. "Black phosphorus nanoflakes/polyaniline hybrid material for high-performance pseudocapacitors." *The Journal of Physical Chemistry C* 121(37): 20532–20538.

Shao, Jundong, Hanhan Xie, Hao Huang, Zhibin Li, Zhengbo Sun, Yanhua Xu, Quanlan Xiao, Xue-Feng Yu, Yuetao Zhao, Han Zhang, Huaiyu Wang, and Paul K. Chu. 2016. "Biodegradable black phosphorus-based nanospheres for in vivo photothermal cancer therapy." *Nature Communications* 7: 12967.

Shen, S. Y., T. S. Zhao, and J. B. Xu. 2010. "Carbon-supported bimetallic PdIr catalysts for ethanol oxidation in alkaline media." *Electrochimica Acta* 55(28): 9179–9184.

Shi, Xuetao, Haixin Chang, Song Chen, Chen Lai, Ali Khademhosseini, and Hongkai Wu. 2012. "Regulating cellular behavior on few-layer reduced graphene oxide films with well-controlled reduction states." *Advanced Functional Materials* 22(4): 751–759.

Song, Ting, Gongchang Zeng, Piyong Zhang, Tingting Wang, Atif Ali, Shaobin Huang, and Heping Zeng. 2019. "3D reticulated carbon nitride materials high-uniformly capture 0D black phosphorus as 3D/0D composites for stable and efficient photocatalytic hydrogen evolution." *Journal of Materials Chemistry A* 7(2): 503–512.

Song, Zeming, Xu Wang, Cheng Lv, Yonghao An, Mengbing Liang, Teng Ma, David He, Ying-Jie Zheng, Shi-Qing Huang, and Hongyu Yu. 2015. "Kirigami-based stretchable lithium-ion batteries." *Scientific Reports* 5: 10988.

Sun, Caixia, Ling Wen, Jianfeng Zeng, Yong Wang, Qiao Sun, Lijuan Deng, Chongjun Zhao, and Zhen Li. 2016. "One-pot solventless preparation of PEGylated black phosphorus nanoparticles for photoacoustic imaging and photothermal therapy of cancer." *Biomaterials* 91: 81–89.

Sun, Jie, Guangyuan Zheng, Hyun-Wook Lee, Nian Liu, Haotian Wang, Hongbin Yao, Wensheng Yang, and Yi Cui. 2014. "Formation of stable phosphorus–carbon bond for enhanced performance in black phosphorus nanoparticle–graphite composite battery anodes." *Nano Letters* 14(8): 4573–4580.

Sun, Li-Qun, Ming-Juan Li, Kai Sun, Shi-Hua Yu, Rong-Shun Wang, and Hai-Ming Xie. 2012. "Electrochemical activity of black phosphorus as an anode material for lithium-ion batteries." *The Journal of Physical Chemistry C* 116(28): 14772–14779.

Sun, Zhengbo, Hanhan Xie, Siying Tang, Xue-Feng Yu, Zhinan Guo, Jundong Shao, Han Zhang, Hao Huang, Huaiyu Wang, and Paul K. Chu. 2015. "Ultrasmall black phosphorus quantum dots: synthesis and use as photothermal agents." *Angewandte Chemie International Edition* 54(39): 11526–11530.

Sun, Zhengbo, Yuetao Zhao, Zhibin Li, Haodong Cui, Yayan Zhou, Weihao Li, Wei Tao, Han Zhang, Huaiyu Wang, and Paul K. Chu. 2017. "TiL4-coordinated black phosphorus quantum dots as an efficient contrast agent for in vivo photoacoustic imaging of cancer." *Small* 13(11): 1602896.

Tang, Qianqiu, Mingming Chen, Gengchao Wang, Hua Bao, and Petr Sáha. 2015. "A facile prestrain-stick-release assembly of stretchable supercapacitors based on highly stretchable and sticky hydrogel electrolyte." *Journal of Power Sources* 284: 400–408.

Tang, Yuxin, Yanyan Zhang, Oleksandr I. Malyi, Nicolas Bucher, Huarong Xia, Shibo Xi, Zhiqiang Zhu, Zhisheng Lv, Wenlong Li, and Jiaqi Wei. 2018. "Identifying the origin and contribution of surface storage in TiO2 (B) nanotube electrode by in situ dynamic valence state monitoring." *Advanced Materials* 30(33): 1802200.

Tang, Yuxin, Yanyan Zhang, Xianhong Rui, Dianpeng Qi, Yifei Luo, Wan Ru Leow, Shi Chen, Jia Guo, Jiaqi Wei, and Wenlong Li. 2016. "Conductive inks based on a lithium titanate nanotube gel for high-rate lithium-ion batteries with customized configuration." *Advanced Materials* 28(8): 1567–1576.

Tao, Wei, Xiaoyuan Ji, Xiaoding Xu, Mohammad Ariful Islam, Zhongjun Li, Si Chen, Phei Er Saw, Han Zhang, Zameer Bharwani, and Zilei Guo. 2017. "Antimonene quantum dots: synthesis and application as near-infrared photothermal agents for effective cancer therapy." *Angewandte Chemie International Edition* 56(39): 11896–11900.

Tao, Wei, Xianbing Zhu, Xinghua Yu, Xiaowei Zeng, Quanlan Xiao, Xudong Zhang, Xiaoyuan Ji, Xusheng Wang, Jinjun Shi, Han Zhang, and Lin Mei. 2017. "Black phosphorus nanosheets as a robust delivery platform for cancer theranostics." *Advanced Materials* 29(1): 1603276.

Viti, Leonardo, Jin Hu, Dominique Coquillat, Wojciech Knap, Alessandro Tredicucci, Antonio Politano, and Miriam Serena Vitiello. 2015. "Black phosphorus terahertz photodetectors." *Advanced Materials* 27(37): 5567–5572.

Walia, Sumeet, Ylias Sabri, Taimur Ahmed, M. Field, Rajesh Ramanathan, Aram Arash, S. Bhargava, Sharath Sriram, Madhu Bhaskaran, and Vipul Bansal. 2016. "Defining the role of humidity in the ambient degradation of few-layer black phosphorus." *2D Materials* 4(1): 1–8.

Wang, Fangfang, Dong Zhai, Chengtie Wu, and Jiang Chang. 2016. "Multifunctional mesoporous bioactive glass/upconversion nanoparticle nanocomposites with strong red emission to monitor drug delivery and stimulate osteogenic differentiation of stem cells." *Nano Research* 9(4): 1193–1208.

Wang, Hua, Bowen Zhu, Wencao Jiang, Yun Yang, Wan Ru Leow, Hong Wang, and Xiaodong Chen. 2014. "A mechanically and electrically self-healing supercapacitor." *Advanced Materials* 26(22): 3638–3643.

Wang, Huaping, Xu-Bing Li, Lei Gao, Hao-Lin Wu, Jie Yang, Le Cai, Tian-Bao Ma, Chen-Ho Tung, Li-Zhu Wu, and Gui Yu. 2018. "Three-dimensional graphene networks with abundant sharp edge sites for efficient electrocatalytic hydrogen evolution." *Angewandte Chemie* 130(1): 198–203.

Wang, Jingping, Chengjun Wu, Peiqi Wu, Xiao Li, Min Zhang, and Jianbo Zhu. 2017. "Polypyrrole capacitance characteristics with different doping ions and thicknesses." *Physical Chemistry Chemical Physics* 19(31): 21165–21173.

Wang, Lili, Joshua A. Jackman, Ee-Lin Tan, Jae Hyeon Park, Michael G. Potroz, Ee Taek Hwang, and Nam-Joon Cho. 2017. "High-performance, flexible electronic skin sensor incorporating natural microcapsule actuators." *Nano Energy* 36: 38–45.

Wang, Lin, Li Huang, Wee Chong Tan, Xuewei Feng, Li Chen, and Kah-Wee Ang. 2018. "Tunable black phosphorus heterojunction transistors for multifunctional optoelectronics." *Nanoscale* 10(29): 14359–14367.

Wang, Renheng, Xinhai Li, Zhixing Wang, and Han Zhang. 2017. "Electrochemical analysis graphite/electrolyte interface in lithium-ion batteries: p-Toluenesulfonyl isocyanate as electrolyte additive." *Nano Energy* 34: 131–140.

Wang, Zhaohui, Petter Tammela, Maria Strømme, and Leif Nyholm. 2015. "Nanocellulose coupled flexible polypyrrole@ graphene oxide composite paper electrodes with high volumetric capacitance." *Nanoscale* 7(8): 3418–3423.

Wei, Qun, and Xihong Peng. 2014. "Superior mechanical flexibility of phosphorene and few-layer black phosphorus." *Applied Physics Letters* 104(25): 251915.

Wei, Teng-Sing, Bok Yeop Ahn, Julia Grotto, and Jennifer A. Lewis. 2018. "3D printing of customized li-ion batteries with thick electrodes." *Advanced Materials* 30(16): 1703027.

Wen, Min, Danni Liu, Yihong Kang, Jiahong Wang, Hao Huang, Jia Li, Paul K. Chu, and Xue-Feng Yu. 2019. "Synthesis of high-quality black phosphorus sponges for all-solid-state supercapacitors." *Materials Horizons* 6(1): 176–181.

Willner, Alan E., Salman Khaleghi, Mohammad Reza Chitgarha, and Omer Faruk Yilmaz. 2013. "All-optical signal processing." *Journal of Lightwave Technology* 32(4): 660–680.

Wood, Joshua D., Spencer A. Wells, Deep Jariwala, Kan-Sheng Chen, EunKyung Cho, Vinod K. Sangwan, Xiaolong Liu, Lincoln J. Lauhon, Tobin J. Marks, and Mark C. Hersam. 2014. "Effective passivation of exfoliated black phosphorus transistors against ambient degradation." *Nano Letters* 14(12): 6964–6970.

Wu, Shuxing, Kwan San Hui, and Kwun Nam Hui. 2018. "2D black phosphorus: from preparation to applications for electrochemical energy storage." *Advanced Science* 5(5): 1700491.

Wu, Tong, Jinchen Fan, Qiaoxia Li, Penghui Shi, Qunjie Xu, and Yulin Min. 2018. "Palladium nanoparticles anchored on anatase titanium dioxide-black phosphorus hybrids with heterointerfaces: highly electroactive and durable catalysts for ethanol electrooxidation." *Advanced Energy Materials* 8(1): 1701799.

Xia, Fengnian, Han Wang, Di Xiao, Madan Dubey, and Ashwin Ramasubramaniam. 2014. "Two-dimensional material nanophotonics." *Nature Photonics* 8(12): 899.

Xiao, Han, Zhong-Shuai Wu, Feng Zhou, Shuanghao Zheng, Dong Sui, Yongsheng Chen, and Xinhe Bao. 2018. "Stretchable tandem micro-supercapacitors with high voltage output and exceptional mechanical robustness." *Energy Storage Materials* 13: 233–240.

Xie, Hanhan, Zhibin Li, Zhengbo Sun, Jundong Shao, Xue-Feng Yu, Zhinan Guo, Jiahong Wang, Quanlan Xiao, Huaiyu Wang, and Qu-Quan Wang. 2016. "Metabolizable ultrathin Bi2Se3 nanosheets in imaging-guided photothermal therapy." *Small* 12(30): 4136–4145.

Xing, Chenyang, Guanghui Jing, Xin Liang, Meng Qiu, Zhongjun Li, Rui Cao, Xiaojing Li, Dianyuan Fan, and Han Zhang. 2017. "Graphene oxide/black phosphorus nanoflake aerogels with robust thermo-stability and significantly enhanced photothermal properties in air." *Nanoscale* 9(24): 8096–8101.

Xing, Chenyang, Shiyou Chen, Meng Qiu, Xin Liang, Quan Liu, Qingshuang Zou, Zhongjun Li, Zhongjian Xie, Dou Wang, and Biqin Dong. 2018. "Conceptually novel black phosphorus/cellulose hydrogels as promising photothermal agents for effective cancer therapy." *Advanced Healthcare Materials* 7(7): 1701510.Xiong, Kun, Qingbo Fan, Tingting Wu, Haishan Shi, Lin Chen, and Minhao Yan. 2017. "Enhanced bovine serum albumin absorption on the N-hydroxysuccinimide activated graphene oxide and its corresponding cell affinity." *Materials Science and Engineering: C* 81: 386–392.

Xu, Gui-Liang, Zonghai Chen, Gui-Ming Zhong, Yuzi Liu, Yong Yang, Tianyuan Ma, Yang Ren, Xiaobing Zuo, Xue-Hang Wu, and Xiaoyi Zhang. 2016. "Nanostructured black phosphorus/Ketjenblack–multiwalled carbon nanotubes composite as high performance anode material for sodium-ion batteries." *Nano Letters* 16(6): 3955–3965.

Xu, Sheng, Yihui Zhang, Jiung Cho, Juhwan Lee, Xian Huang, Lin Jia, Jonathan A. Fan, Yewang Su, Jessica Su, and Huigang Zhang. 2013. "Stretchable batteries with self-similar serpentine interconnects and integrated wireless recharging systems." *Nature Communications* 4: 1543.

Yang, Bowen, Junhui Yin, Yu Chen, Shanshan Pan, Heliang Yao, Youshui Gao, and Jianlin Shi. 2018. "2D-black-phosphorus-reinforced 3D-printed scaffolds: a stepwise counter-measure for osteosarcoma." *Advanced Materials* 30(10): 1705611.

Yang, Hui, Wan Ru Leow, Ting Wang, Juan Wang, Jiancan Yu, Ke He, Dianpeng Qi, Changjin Wan, and Xiaodong Chen. 2017. "3D printed photoresponsive devices based on shape memory composites." *Advanced Materials* 29(33): 1701627.

Youngblood, Nathan, Che Chen, Steven J. Koester, and Mo Li. 2015. "Waveguide-integrated black phosphorus photodetector with high responsivity and low dark current." *Nature Photonics* 9(4): 247.

Youngblood, Nathan, Ruoming Peng, Andrei Nemilentsau, Tony Low, and Mo Li. 2016. "Layer-tunable third-harmonic generation in multilayer black phosphorus." *ACS Photonics* 4(1): 8–14.

Yuan, Hongtao, Xiaoge Liu, Farzaneh Afshinmanesh, Wei Li, Gang Xu, Jie Sun, Biao Lian, Alberto G. Curto, Guojun Ye, and Yasuyuki Hikita. 2015. "Polarization-sensitive broadband photodetector using a black phosphorus vertical p–n junction." *Nature Nanotechnology* 10(8): 707.

Yun, Qinbai, Qipeng Lu, Xiao Zhang, Chaoliang Tan, and Hua Zhang. 2018. "Three-dimensional architectures constructed from transition-metal dichalcogenide nanomaterials for electrochemical energy storage and conversion." *Angewandte Chemie International Edition* 57(3): 626–646.

Zhang, Jintao, Zhen Li, and Xiong Wen Lou. 2017. "A freestanding selenium disulfide cathode based on cobalt disulfide-decorated multichannel carbon fibers with enhanced lithium storage performance." *Angewandte Chemie International Edition* 56(45): 14107–14112.

Zhang, Panpan, Feng Zhu, Faxing Wang, Jinhui Wang, Renhao Dong, Xiaodong Zhuang, Oliver G. Schmidt, and Xinliang Feng. 2017. "Stimulus-responsive micro-supercapacitors with ultrahigh energy density and reversible electrochromic window." *Advanced Materials* 29(7): 1604491.

Zhang, Qiancheng, Xiaohu Yang, Peng Li, Guoyou Huang, Shangsheng Feng, Cheng Shen, Bin Han, Xiaohui Zhang, Feng Jin, and Feng Xu. 2015. "Bioinspired engineering of honeycomb structure: using nature to inspire human innovation." *Progress in Materials Science* 74: 332–400.

Zhang, Qiangqiang, Yu Wang, Baoqiang Zhang, Keren Zhao, Pingge He, and Boyun Huang. 2018. "3D superelastic graphene aerogel-nanosheet hybrid hierarchical nanostructures as high-performance supercapacitor electrodes." *Carbon* 127: 449–458.

Zhang, Yupu, Lili Wang, Hao Xu, Junming Cao, Duo Chen, and Wei Han. 2020. "3D chemical cross-linking structure of black phosphorus@ CNTs hybrid as a promising anode material for lithium ion batteries." *Advanced Functional Materials* 30(12): 1909372.

Zhao, Yuetao, Huaiyu Wang, Hao Huang, Quanlan Xiao, Yanhua Xu, Zhinan Guo, Hanhan Xie, Jundong Shao, Zhengbo Sun, and Weijia Han. 2016. "Surface coordination of black phosphorus for robust air and water stability." *Angewandte Chemie International Edition* 55(16): 5003–5007.

Zheng, Shijun, Enxiu Wu, Zhihong Feng, Rao Zhang, Yuan Xie, Yuanyuan Yu, Rui Zhang, Quanning Li, Jing Liu, and Wei Pang. 2018. "Acoustically enhanced photodetection by a black phosphorus–MoS 2 van der Waals heterojunction p–n diode." *Nanoscale* 10(21): 10148–10153.

Zhou, Qionghua, Qian Chen, Yilong Tong, and Jinlan Wang. 2016. "Light-induced ambient degradation of few-layer black phosphorus: mechanism and protection." *Angewandte Chemie International Edition* 55(38): 11437–11441.

Zhou, Ye, Maoxian Zhang, Zhinan Guo, Lili Miao, Su-Ting Han, Ziya Wang, Xiuwen Zhang, Han Zhang, and Zhengchun Peng. 2017. "Recent advances in black phosphorus-based photonics, electronics, sensors and energy devices." *Materials Horizons* 4(6): 997–1019.

Zhu, Mingshan, Sooyeon Kim, Liang Mao, Mamoru Fujitsuka, Junying Zhang, Xinchen Wang, and Tetsuro Majima. 2017. "Metal-free photocatalyst for H2 evolution in visible to near-infrared region: black phosphorus/graphitic carbon nitride." *Journal of the American Chemical Society* 139(37): 13234–13242.

Zhu, Mingshan, Zhichao Sun, Mamoru Fujitsuka, and Tetsuro Majima. 2018. "Z-scheme photocatalytic water splitting on a 2D heterostructure of black phosphorus/bismuth vanadate using visible light." *Angewandte Chemie International Edition* 57(8): 2160–2164.

Zhu, Xianjun, Taiming Zhang, Daochuan Jiang, Hengli Duan, Zijun Sun, Mengmeng Zhang, Hongchang Jin, Runnan Guan, Yajuan Liu, and Muqing Chen. 2018. "Stabilizing black phosphorus nanosheets via edge-selective bonding of sacrificial C 60 molecules." *Nature Communications* 9(1): 1–9.

Zhu, Xianjun, Taiming Zhang, Zijun Sun, Huanlin Chen, Jian Guan, Xiang Chen, Hengxing Ji, Pingwu Du, and Shangfeng Yang. 2017. "Black phosphorus revisited: a missing metal-free elemental photocatalyst for visible light hydrogen evolution." *Advanced Materials* 29(17): 1605776.

Zhu, Zhiqiang, Yuxin Tang, Zhisheng Lv, Jiaqi Wei, Yanyan Zhang, Renheng Wang, Wei Zhang, Huarong Xia, Mingzheng Ge, and Xiaodong Chen. 2018. "Fluoroethylene carbonate enabling a robust lif-rich solid electrolyte interphase to enhance the stability of the mos2 anode for lithium-ion storage." *Angewandte Chemie International Edition* 57(14): 3656–3660.

7 Brief Summary and Future Directions

7.1 BRIEF SUMMARY

Different synthetic procedures (top-down and bottom-up approaches) containing properties of black phosphorus (BP), composite of BP with polymers and other 2D layered material, comparison of BP with other 2D layered material, and construction of 3D structures of BP and their applications in the field of energy and biomedical have been reviewed. It was found that 2D BP nanomaterials including BP nanotubes, bulk and monolayer, and quantum dot BP possess outstanding optical, biocompatible, thermal, mechanical and electrical properties besides other characteristics such as nontoxicity and chemical and thermal stability values. Different 3D structures of BP compounds are thus attractive candidates for multiple technological and industrial applications including optoelectronics, energy storage, and biocompatible material for drug delivery applications. Although there is a successful integration of 2D BP as a semiconductor material compared with another counterpart including Transition metal dichalcogenide (TMDCs) and graphene, for the final application of BP structures at the nano-scale, however, there is quite a shortage of such methods in large-scale production of BP structures due to various uncontrolled defects, the presence of secondary phases/disorder, as well as the impurities. The growth of a 3D BP architecture for practical applications is highly desired. Expecting to give even better presentation for practical applications in the sphere of energy and pharmaceuticals due to the 3D structure of 2D nanomaterials rather than 2D architecture. BP can be developed as an alternative material for minimizing the hurdles when designing ultrathin large-area membranes for channels in the silicon semiconductor industry for the preparation of 3D hybrid structures (Akahama, Endo, and Narita 1983; Doganov et al. 2015). The BP laminated layer/membrane-based 3D hybrid systems having the great van der Waals (vdWs) interaction for other 2D layered material, quantum dots, polymers, and small organic molecules, and act as a good host/hybrid structure with enhanced electrical, mechanical, and biodegradable properties. Also, on the other side, improvement is still required in synthetic protocol and assembly to 3D macrostructures of BP for potential technological applications. Recently, beyond the 3D hybrid structure, other 3D structures that are under consideration and active area of research are 3D macrostructures of BP. It is analyzed that the 3D macroscopic systems of 2D materials are porous in nature, so can be used as electrodes for batteries/supercapacitor applications (Chen 2013; Wang

DOI: 10.1201/9781003217145-7

et al. 2010; Guo, Song, and Chen 2009; Hou et al. 2011; Simon and Gogotsi 2008; Wang, Zhang, and Zhang 2012; Zhang and Zhao 2009). Moreover, the pores can host various other materials, including metals, organic molecules, polymers, etc. (Dong et al. 2012a, 2012b; Yong et al. 2012; Sattayasamitsathit et al. 2013; Maiyalagan et al. 2012) resulting in new functionalities and applications. 3D macrostructures of black phosphorus (BP) will contain the fundamental characteristics of BP containing the high mobility, tunable narrow bandgap, and mechanical strength and contain new functionalities due to their porous structure.

These pores facilitate ion transport and diffusion, and the interconnecting of the 2D nanosheets in the formation of 3D structure provides abundant paths for electron transport. These attributes are key for well-organized energy conversion and storage materials. From the above discussion, we can conclude that 2D nanosheets of BP materials may have applications in electronic devices, but for energy conversion/storage and other biomedical applications where porous large scale structures are required, 3D macrostructures of BP can come in handy (Lv et al. 2018; Wen et al. 2019; Yang et al. 2018; Xing et al. 2018). However, BP nanosheets' sensitivity to air water and oxidants is not very suitable for its practical applications and there is much more to do to improve the long-term stability properties of BP-based structures. In addition, there is also a need for improvement in the construction of 3D BP. Even though many different approaches have been established for the synthesis of 3D BP architecture, further efforts are required for large-scale employment in nanotechnological applications.

7.2   FUTURE DIRECTIONS

The synthetic challenges cannot deter the scientist around the world from further exploring the latest 2D materials based on BP. In electronic, electrochemical, and optical sensor modules, BP has been widely used as electrode modifiers, using their different features. However, 2D BP is not appropriate for its practical use in processing techniques. Far more needs to be done to boost the long-term stability of BP-based manufactured heterostructures. Furthermore, after applying the various techniques, such as spinning coating, drop-casting, spray coating, etc., for technologically based potential applications, the role of solvency and other parameters for improving synthetic protocol and assembling BP-founded heterostructures needs to be investigated.

Overall, BP-based 3D structures are expected to advance at a developing rate in less than ten years for biomedical, sensing, and other potential applications. Besides the biomedical and potential application of 3D architecture of BP, we can suggest the effective sensing applications. Recently, BP is currently undergoing analytical sensing applications, researchers are expected to explore the full potential of this new class of sensor materials. In pollutant removal (such as removal of heavy metals and dyes from water) and environmental remediation, use of based materials is required.

We hope to develop sensing applications when navigating through different structures such as the quantum dots, nanoflakes, monolayers, or few layers of BP. In addition, this study will guide materials scientists and engineers to contribute to this progress through studies and future directions. We also hope that this book will allow researchers in physics, chemistry, and engineering to consider the BP options easily.

7.3 CONCLUSIONS

2D BP nanomaterials like BP nanotubes, bulk and monolayers, and quantum dot BP, beside other characteristics including non-toxicity and chemical and thermal stability qualities, have outstanding optical, bio-compatible, thermal, mechanical, and electrical properties. Thus, various 3D compound structures are desirable candidates for a wide variety of scientific and industrial uses, from optoelectronics, energy storage and biologically compatible materials for drug delivery applications.

7.4 ACKNOWLEDGMENTS

The research was partially supported by financial support from the Science and Technology Development Fund (Nos. 007/2017/A1 and 132/2017/A3), Macao Special Administration Region (SAR), China, and National Natural Science Fund (Grant Nos. 61875138, 61435010, 61805147, and 6181101252), and Science and Technology Innovation Commission of Shenzhen (KQTD20150324 16270385, JCYJ20150625103619275, JCYJ20180305125141661, and JCYJ2017081 1093453105). The authors also acknowledge the support from the Instrumental Analysis Center of Shenzhen University (Xili Campus)

REFERENCES

Akahama, Yuichi, Shoichi Endo, and Shin-ichiro Narita. 1983. "Electrical properties of black phosphorus single crystals." *Journal of the Physical Society of Japan* 52(6): 2148–2155.
Chen, Jiajun. 2013. "Recent progress in advanced materials for lithium ion batteries." *Materials* 6(1): 156–183.
Doganov, Rostislav A., Eoin C. T. O'Farrell, Steven P. Koenig, Yuting Yeo, Angelo Ziletti, Alexandra Carvalho, David K. Campbell, David F. Coker, Kenji Watanabe, Takashi Taniguchi, Antonio H. Castro Neto, and Barbaros Özyilmaz. 2015. "Transport properties of pristine few-layer black phosphorus by van der Waals passivation in an inert atmosphere." *Nature Communications* 6: 6647.
Dong, Xiaochen, Jingxia Wang, Jing Wang, Mary B. Chan-Park, Xingao Li, Lianhui Wang, Wei Huang, and Peng Chen. 2012. "Supercapacitor electrode based on three-dimensional graphene–polyaniline hybrid." *Materials Chemistry and Physics* 134(2–3): 576–580.
Guo, Peng, Huaihe Song, and Xiaohong Chen. 2009. "Electrochemical performance of graphene nanosheets as anode material for lithium-ion batteries." *Electrochemistry Communications* 11(6): 1320–1324.
Hou, Junbo, Yuyan Shao, Michael W. Ellis, Robert B. Moore, and Baolian Yi. 2011. "Graphene-based electrochemical energy conversion and storage: fuel cells, supercapacitors and lithium ion batteries." *Physical Chemistry Chemical Physics* 13(34): 15384–15402.
Lv, Zhisheng, Yuxin Tang, Zhiqiang Zhu, Jiaqi Wei, Wenlong Li, Huarong Xia, Ying Jiang, Zhiyuan Liu, Yifei Luo, and Xiang Ge. 2018. "Honeycomb-lantern-inspired 3D stretchable supercapacitors with enhanced specific areal capacitance." *Advanced Materials* 30(50): 1805468.

Maiyalagan, T., X. C. Dong, P. Chen, and X. Wang. 2012. "Electrodeposited Pt on three-dimensional interconnected graphene as a free-standing electrode for fuel cell application." *Journal of Materials Chemistry* 22(12): 5286–5290.

Sattayasamitsathit, S., Y. E. Gu, K. Kaufmann, W. Z. Jia, X. Y. Xiao, M. Rodriguez, S. Minteer, J. Cha, D. B. Burckel, C. M. Wang, R. Polsky, and J. Wang. 2013. "Highly ordered multilayered 3D graphene decorated with metal nanoparticles." *Journal of Materials Chemistry A* 1(5): 1639–1645.

Simon, Patrice, and Yury Gogotsi. 2008. "Materials for electrochemical capacitors." *Nature Materials* 7(11): 845–854.

Wang, Guoping, Lei Zhang, and Jiujun Zhang. 2012. "A review of electrode materials for electrochemical supercapacitors." *Chemical Society Reviews* 41(2): 797–828.

Wang, Hailiang, Li-Feng Cui, Yuan Yang, Hernan Sanchez Casalongue, Joshua Tucker Robinson, Yongye Liang, Yi Cui, and Hongjie Dai. 2010. "Mn3O4–graphene hybrid as a high-capacity anode material for lithium ion batteries." *Journal of the American Chemical Society* 132(40): 13978–13980.

Wen, Min, Danni Liu, Yihong Kang, Jiahong Wang, Hao Huang, Jia Li, Paul K Chu, and Xue-Feng Yu. 2019. "Synthesis of high-quality black phosphorus sponges for all-solid-state supercapacitors." *Materials Horizons* 6(1): 176–181.

Xing, Chenyang, Shiyou Chen, Meng Qiu, Xin Liang, Quan Liu, Qingshuang Zou, Zhongjun Li, Zhongjian Xie, Dou Wang, and Biqin Dong. 2018. "Conceptually novel black phosphorus/cellulose hydrogels as promising photothermal agents for effective cancer therapy." *Advanced Healthcare Materials* 7(7): 1701510.

Yang, Bowen, Junhui Yin, Yu Chen, Shanshan Pan, Heliang Yao, Youshui Gao, and Jianlin Shi. 2018. "2D-black-phosphorus-reinforced 3D-printed scaffolds: a stepwise counter-measure for osteosarcoma." *Advanced Materials* 30(10): 1705611.

Yong, Yang-Chun, Xiao-Chen Dong, Mary B. Chan-Park, Hao Song, and Peng Chen. 2012. "Macroporous and monolithic anode based on polyaniline hybridized three-dimensional graphene for high-performance microbial fuel cells." *ACS Nano* 6(3): 2394–2400.

Zhang, Li Li, and X. S. Zhao. 2009. "Carbon-based materials as supercapacitor electrodes." *Chemical Society Reviews* 38(9): 2520–2531.

Index